M000235832

CORN AMONG THE INDIANS
OF THE UPPER MISSOURI

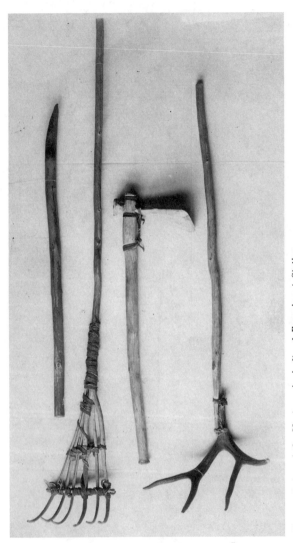

By permission of the Montana Agricultural Experiment Station

SET OF HIDATSA AGRICULTURAL IMPLEMENTS

Digging stick, willow rack, iron hoe, and antler rake

CORN AMONG THE INDIANS OF THE UPPER MISSOURI

BY
GEORGE F. WILL
AND
GEORGE E. HYDE

UNIVERSITY OF NEBRASKA PRESS
LINCOLN AND LONDON

Manufactured in the United States of America

⊗

First Nebraska paperback printing: 1964

Library of Congress Cataloging-in-Publication Data
Will, George F. (George Francis), 1884–1955.
Corn among the Indians of the Upper Missouri / by George F.
Will and George E. Hyde.
p. cm.
Originally published: Saint Louis, Mo.: W. H. Miner Co.,
1917.
Includes bibliographical references and index.
ISBN 0-8032-9826-9 (pbk.: alk. paper)
1. Indians of North America—Agriculture—Missouri River
Valley. 2. Corn. I. Hyde, George E., 1882–1968. II. Title.
E98.A3.W6 2002
630′.89′975—dc21   2002028521

DEDICATION

With the permission of Mr. Hyde I dedicate this volume to my father, Oscar H. Will, who in 1881 first perceived the value of the native varieties of corn from the Upper Missouri Valley, and who began at that time the work of selecting and breeding from them, to the lasting benefit of the farmers of the Northwest.

GEORGE F. WILL

CONTENTS

ILLUSTRATIONS

CORN AMONG THE INDIANS
OF THE UPPER MISSOURI

ACKNOWLEDGMENTS

In preparing this study of the agriculture, and more particularly the corn culture, of the Indian tribes of the Upper Missouri region we have of course drawn items from very many sources, and we are indebted to many persons for assistance of one sort or another. In working over the material we have conducted rather extensive experiments in raising corn from the different tribes in order to gain more complete information on the varieties.

For a deep and encouraging interest in the work and much material aid in the matter of furnishing seed of a great many kinds of corn we are especially indebted to Mr. M. L. Wilson of the Montana Agricultural College, State Leader of County Agents, who has also furnished some interesting data, from which we have drawn, in the Montana Experiment Station Bulletin No. 107.

For seed and information supplied we are also greatly indebted to Dr. Melvin R. Gilmore, Mr. H. C. Fish, of the Minot, N. D., Normal School, to Mr. Thomas Cory, of the Moose Mountain Reserve, Carlysle, Sask., to Mr. James McDonald, of the Pipestone Reserve, Griswold, Manitoba, Mr.

R. P. Stanion, Superintendent, Otoe Agency, Otoe, Okla., Mr. J. H. Johnson, Superintendent, Sac and Fox Indian School, Stroud, Okla., and to George Bent of the Cheyennes, Joseph Springer of the Iowas, and James Murie of the Pawnees.

Mr. Will's special acknowledgements are due to Dr. Gilbert L. Wilson of Minneapolis, who has spent much time and enthusiasm in collecting data on Hidatsa agriculture, for kindness in permitting a reading of his notes and for opportunities for profitable comparison of material on the Hidatsas; to Scattered Corn, daughter of the last Mandan corn priest, and to her son, James Holding Eagle, for a great deal of assistance in everything concerning Mandan agriculture and for a deep interest in having the subject properly understood and presented; to Dr. Melvin R. Gilmore of the North Dakota Historical Society, formerly of the Nebraska Historical Society, for seed of many varieties of corn and for much profitable interchange of thought and knowledge on the subject of Indian agriculture; to various members of the staff of the American Museum of Natural History of New York, for specimens of corn and for lending encouragement to the study; to Professor R. B. Dixon of Harvard University for much encouraging interest and for opportune and kindly advice; and to Dr. C. L. Hall, the veteran missionary of the Fort Berthold Reservation, and to his family, for many favors accorded during every visit to the reservation, for data on the corn of the three Berthold tribes, and especially for information on basket-making and the acquisition of a number of baskets.

Mr. Hyde's acknowledgements are also due to Dr. Gilmore, for seed and information supplied on many occasions; and to Mr. Duncan C. Scott, of the Canadian Indian Office, for assistance in obtaining information on Canadian Indian corn; to Mr. Thomas Cory of the Moose Mountain Reserve, for information on Assiniboin agriculture and for seed furnished; to Mr. James McDonald of the Griswold Agency, for similar information on the Refugee-Sioux and for seed from that tribe; to Mr. George Bird Grinnell of New York City, for much assistance and advice and for some very interesting notes on Cheyenne agriculture; to Mr. G. N. Collins of the Bureau of Plant Industry, Department of Agriculture, Washington, D.C., for information and for assistance rendered on several occasions; and to George Bent of the Cheyennes, for information and for some ears of Wichita corn.

No list of the printed authorities consulted has been compiled, but the sources from which material of this class has been drawn are always indicated in the text or in the footnotes. The Coues' edition of Lewis and Clark was used except in a few instances in which the Thwaites' edition of the original journals was consulted. The quotations from Long, Maximilian, Brackenridge, and Bradbury are all taken from the Thwaites, *Early Western Travels*.

INTRODUCTION

Corn, as every one knows but most of us often forget, is a gift to us from the Indian race. In the early period of colonization, our ancestors along the Atlantic coast were glad and even thankful to take the Indian's corn and to learn from him the methods of growing and handling the crop. With the first knowledge of corn and its culture received from the tribes near the coast from New England to Virginia and Carolina, however, the American farmer has felt, generally, that there was nothing further to be learned from the Indians about corn. The pioneer too often failed to realize that each new region settled presented new conditions of soil and climate and that the best way to learn how to meet and overcome these conditions was to study the methods of the local tribes, who often had been growing corn in that particular region for two or three hundred years, and who during this long period of time had learned from hard experience the varieties of corn and the cultural methods best suited to the local conditions.

The first white settlers in the Upper Missouri country failed to understand these facts. For a

long time the only whites in all this region were
the fur traders and their followers, the free trappers, and the United States Indian agents. None
of these men were engaged in agriculture. The
traders for the most part thought of corn only as
something to be bought from the Indians, while
their followers, who often married into the tribes,
were usually content to let their wives raise what
corn was needed without themselves giving much
attention to the matter.

We find some of the Indian agents experiment-
ing with corn at a rather early date; but for the
most part the corn they tried out was from the
east; the native varieties were discarded, usually
without investigation as to their usefulness, and
we often find the Indian corn, even as far up as
Fort Berthold in North Dakota, mentioned dis-
paragingly in the agents' reports. In the more
northern and arid sections however the fact soon
became evident that eastern varieties of corn
could not be depended upon to make a crop, and
thereafter some of the agents became really in-
terested in the hardier types of native corn. The
agent for the Red Cloud Sioux in his report for
1873 referred to the Ree corn as a hardy acclimated variety and asked to be supplied with seed
of this sort. Soon after that date we find the
Indian Office sending out Ree seed to be tested
at several of the northern agencies. Some Ree

corn was sent to the Fort Peck Agency, Montana, in 1878 (apparently also to the Crow Agency in Montana) and proved very satisfactory.

As to whether the corn of the tribes of the lower Missouri Valley was extensively grown by the pioneer white settlers, the records are silent; but it seems probable that these native varieties were used, to some extent, during the first period of settlement. Sturtevant in his list of varieties grown by the whites (1884) mentions both an Omaha blue corn and a Mandan "squaw corn," the former variety being grown as far east as Illinois; while an old settler in Kansas speaks of a very early variety, called yellow maiden corn, seemingly a native corn, which he implies was popular among the pioneer settlers in the lower valley of Kansas River. We hear occasionally of blue corn, white flour corn, and mixed corn of numerous types, all usually lumped together as "squaw corn," and always as pioneer varieties.

It has been the history of western settlement, however, that as soon as the pioneer conditions were overcome in a new region and it was demonstrated to the settlers' satisfaction that the new country was not so very different from the old home back east, at once the desire arose to cultivate the same crops and the same varieties of corn that they had formerly raised in the east, in order to form a comparison of conditions in their new

home with those in the old.   Such attempts to ac-
climatize corn from the east or south in newly
settled regions often proved successful and usually led to the immediate abandonment of the native
corns which the first settlers in the new country
had of necessity made use of.   But in the north
and in the arid west these attempts often failed.
In New England, although experiments with dent
corn have been made from time to time, the flint
corns obtained by the first settlers from the Indians are still the main dependence.   The well
known and widely diffused King Philip corn is
one of the surviving varieties of the pioneer New
England flints.   This permanency of the original
Indian varieties is true also to a considerable extent in New York.   Farther south the dent corns
were received from the Indians and, being prolific
in the south and in all of the milder regions, these
varieties soon came to be the staple type in the
southern and central or Corn Belt states.

As the frontier moved westward and northward from the milder central states, and as the
settlers became established and corn selection and
breeding began to claim attention, the desire to
emulate the achievements of ''back home'' in corn
raising made it almost inevitable that the dent
corn should be chosen for cultivation by the bulk
of the people.   The more the dent corn area en-
larged its bounds the more readily did its acquired

momentum push it farther and farther, frequently so that no other types of corn were ever tried in many of the newly settled localities.

In this way Iowa, Nebraska, all the states lower down the river, and even southeastern South Dakota very rapidly passed through whatever pioneer period there was of growing "squaw corn," and became part of the dent corn area. Southern Minnesota did not abandon the hardy Indian flints as early — perhaps because of the Scandinavian population who had never raised corn "back home" and who also were naturally conservative. Robinson says (*Early Economic Conditions and the Development of Agriculture in Minnesota,* 1915, p. 176) that up to 1899 corn was only raised in the southern tier of counties and was mostly flint and squaw corn.

It is obvious from the above that the native corns of the Indians along the lower course of the Missouri, however valuable they may have proven had the attention of breeders been turned to them, were neglected, and allowed in many cases to degenerate and often to disappear, without arousing the slightest interest or consideration from the agricultural investigators of the region. It seems very possible that qualities of value may have been thus lost.

As the frontier moved up the Missouri valley into South Dakota, however, the triumphal progress of the dent corns began to slacken. As the latitude and altitude both increased and the rainfall became less, it grew to be a much longer and more difficult task to acclimate dent varieties of corn. In this region also the flint and flour corns of the Rees and Mandans had acquired some reputation all up and down the river and even in the writings of the time, for their extreme hardiness. Ree corn, talked of by all the tribes, had been officially distributed on many of the reservations. Even in fur trading times corn from the Mandans was taken to the Red River posts. In fact it is probable that the corn of the northern Minnesota Chippewas was derived from this source, so like is it in general character and appearance.

The early traders, the wood hawks, and such of the nomadic frontiersmen as married Indian women and settled down, all raised the corn of the Indians; and when the first real settlement of the region began in the seventies, bringing with it a real farming population, many of these oldtimers took up pieces of land along with the newcomers. Very soon there arose a demand for corn among these pioneer farmers from the east and for a time the mixed Indian corn was adopted by all of them.

Some few of the new settlers saw the real value and probable importance of the hardy native corn and started the work of improving it; and all of the better farmers soon established improved

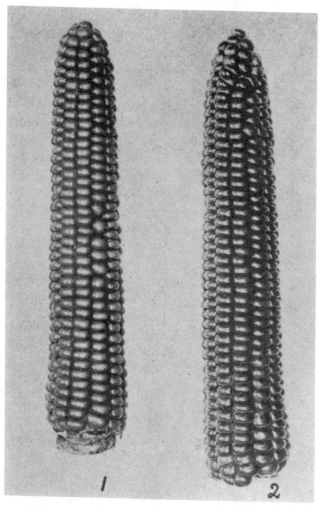

1. Dakota white flint 2. Gehu yellow flint

strains of the mixed flints. In 1882 Mr. O. H. Will of Bismarck, N. D., procured seed of the mixed flint from the Fort Berthold Indians, and Mr. Alphonso Boley, of Mandan, N. D., also secured some at about the same time. From these acquisitions both began the selection and improvement of a pure white flint. Mr. Boley selected for size and large number of rows and produced a large-cobbed corn which shells a rather low percentage and which is slow in drying out in the fall; it has never been very popular, and has not been raised outside of the Missouri Valley in North Dakota.

Mr. Will selected eight and ten rowed ears, long but with a small cob which dries out early. This corn, called Dakota White Flint, was first distributed in 1886, and its spread over the newly settled regions of the Northwest was immediate. It is described as follows in Bulletin No. 107 of the Montana Experiment Station: "Early flint; ears six to ten inches long, and gently tapering; eight to twelve rows of kernels; kernels white, small and blocky; cob white; stalks from twenty-eight to forty-eight inches high according to season; ears borne from four to ten inches from the ground; numerous suckers and fine leaves."

At about the time of the first distribution of the Dakota Flint, Mr. L. D. Judkins of Bismarck raised some and crossed it with Mercer Flint, a later eastern flint, and from this cross was produced the best known corn of Mandan origin, the Gehu Yellow Flint.

In addition to the two above varieties, this Montana bulletin, which is the first publication to give a satisfactory description of the northwestern varieties of corn, names the following varieties, derived from the old Mandan and Ree Indian corns:

Burleigh County Mixed. Catalogued by Mr. Will in 1887. An improved mixed flint of Mandan origin . . . of quite high yield and good quality.

Mixed Indian Corn. Squaw corn, flint and starch corn mixed.

Fort Peck Indian Corn. Mixed early flint. This strain is of Mandan origin, but has been grown on the Fort Peck and Fort Belknap Indian reservations for over thirty years, and is now earlier in season than the modern Mandan corn.

Beal Corn. Much raised in Burleigh, Emmons, and contiguous counties in North Dakota. It is a cross between the mixed flint and some unknown dent, and is very hardy and a heavy yielder.

Ivory King. An improved variety of Indian white flour corn from unknown sources.

Ree Corn. Mixed flour corn, grown in South Dakota.

All of these varieties show the exceptionally hardy characteristics of the Mandan and Arikara

corn, ripen very early, and, under favorable conditions, will yield as high as fifty to seventy bushels
per acre. The Dakota White Flint, of pure Mandan blood, on a Northern Pacific Railway demonstration farm near Elgin, N. D., gave a yield of
seventy bushels per acre in 1914. It has yielded
over forty bushels per acre in eastern Montana,
and often reaches that figure for very large fields
in North Dakota; this variety has also ripened a
good crop at Bozeman, Mont., at an elevation of
6,000 feet. The Gehu has ripened well and yielded
fifty bushels per acre in the Flathead section of
Montana, under the shadow of snow-capped mountains.

For some fifteen or twenty years these hardy
types of corn were the only ones grown in the
Northwest, and many a struggling homesteader
in the lean years owed his home and the founda-
tion of his success to them; when wheat failed and
there was no market for other grains, flint corn
and a few pigs and cattle invariably tided him
over the winter.

Meanwhile, however, the improvement and
breeding of corn for this region remained at a
standstill, and the crossing experiments and selecting that should have been done in those years
remain for the most part undone, even yet.

An unusually hardy and early red dent was
brought into the Missouri Valley in North Dakota

along toward the beginning of the second period of settlement, in the late nineties, and soon became fairly well acclimated. This corn, called Northwestern Dent, is really a semi-dent, and shows the characteristics of the Indian types of corn both in appearance and ability to withstand severe conditions. As soon as this corn became distributed and generally known the period of emulation of the older corn country was open and almost everyone rushed to the trying out and acclimating of various dent corns.

In the rigorous process of selection under climatic conditions the hardiest dent types were soon all that survived, and it has become more and more apparent that, outside of the few localities with exceptionally heavy soil and fairly low altitude such as the southeastern part of the Red River Valley, as the dents become hardy and acclimated the ears become smaller and the yield less. The Northwestern Dent is the only sort of dent that has kept a fairly constant yield, and, while it surpasses perhaps in amount of silage, yet it hardly comes up to the native flints in yield of grain.

Some few of the most conservative of the old settlers did not take up with the new dents, and gradually their neighbors have learned that the dents are being consistently outyielded by the old native varieties, except upon the heaviest and richest soils. This has been grudgingly admitted, especially by the agricultural experts and authorities, but within the past three years there has been a steady shift in the current of agricultural sentiment in the northwestern states and the demand for the native flints is steadily and rapidly increasing. The common sense report on these varieties, in Bulletin No. 107 of the Montana Experiment Station, by Professor Alfred Atkinson and M. L. Wilson has done much to bring this change about. The common cry of the opponents of the native flints is that corn of this type grows so low that the ears are very hard to gather. This is the only valid argument against these flints; and surely as their superiority is continually more and more clearly demonstrated, machinery to handle the crop properly will be invented.

Several years ago, at about the time when it began to be realized that dent corn could not be generally successful in the Northwest, interest in the old native varieties was revived and six or seven men, working independently, took up anew the study of the Indian corn of the Missouri Valley. At first it was supposed that many of these varieties had been lost and that most of the others had been permitted to degenerate into the condition of mere squaw corn, but as the work progressed it was learned that a surprising number of the old varieties still existed in pure or almost pure strains. Thus of thirteen varieties said to have been grown by the Mandans in early times, all except four have been found, and most of these varieties are still quite pure strains today. Among the Pawnees nine varieties were found, practically all pure strains, and among the Omahas about the same number, half of them pure strains, the rest rather badly mixed. Altogether some fifty varieties have been found among the tribes that formerly practised agriculture in the Missouri Valley, not including any of the Sioux.

In the following pages the present authors intend to describe these newly discovered varieties of native corn; and to give some account of the agricultural methods of the Upper Missouri Indians, of their manner of harvesting and storing the crop, of the ways in which they prepared corn for food, of their traditions relating to the origin of corn, and of their corn rites. This work should have been undertaken fifty years ago when a great deal of material, now lost, was still available. At this late day the task has been fraught with many difficulties, some of which have proved insurmountable; and as it was not possible for the authors to make personal visits to all of the tribes, much valuable material has been missed — material which certainly still exists, but which will be lost with the death of the older Indians, who alone know of these things.

The work of collecting seed of the old Indian varieties of corn has been very successful; nearly all of the sorts formerly grown by the tribes along the Missouri, from the Platte northward, have been recovered, experimental plantings have been made, and the seed has been rather widely distributed among corn-breeders.

I — THE UPPER MISSOURI INDIANS

1. MIGRATIONS AND EARLY HISTORY. 2. THE EARTH-
LODGE VILLAGE. 3. AGRICULTURE

1. *Migrations and early history*

In the days of the early fur traders the Upper Missouri country was usually considered to be the region along the Missouri River from the mouth of the Platte up to the Rocky Mountains and embracing a large area of country on both sides of the river. This definition is practically the one we shall follow; for although the agricultural tribes occupied only the immediate valley of the Missouri, and of the lower Platte, yet their influence was felt over large tracts of country bordering the river valley. The Missouri, with its tributaries, was the backbone of the whole region, and played a prominent part in the life of nearly all of the tribes, even the more distant tribes in the Plains making occasional journeys to the Indian villages on the Missouri, to barter and visit.

In this paper we shall deal with the agricultural tribes of the Upper Missouri area, and the hunter tribes will be mentioned only incidentally.

These agricultural tribes of the Upper Missouri belonged to two linguistic stocks: the Caddoan and Siouan.[1]

---

[1] Two Algonquian tribes, the Cheyennes and Arapahoes, planted in the Missouri Valley in South Dakota for many years in the

The Caddoan stock was represented by the four tribes of Pawnees, living on the lower Platte in Nebraska, and by the Arikaras, close kindred of the Pawnees, who, when first visited by the French in 1743, lived in several villages on the Missouri River, near the present Pierre, South Dakota. Like the Pawnees, the Arikaras were a numerous people, divided into several tribes, before pestilence and war reduced them to a single weak village late in the nineteenth century. These Caddoan peoples came from the southwest at a very early date, and their advent into the Upper Missouri country probably antedates that of the first Siouan arrivals. The Pawnees, and evidently the Arikaras also, had already established themselves in northern Kansas and southern Nebraska when Coronado visited the tribes on Kansas River (1541), and by the year 1723 the Pawnees had moved up to the Platte Valley, where they remained for one hundred and fifty years, while the Arikaras, about the same time, had gone on up the Missouri into South Dakota, where they soon after came into contact with the Mandans, with whom they are still living today.

The Mandans were evidently the first Siouan tribe to reach the Upper Missouri. They seem to have come from the east, perhaps from the Minnesota region, but all knowledge of an older home, outside of the Missouri Valley, has been lost. Their coming dates back at least three hundred years. Their earliest traditions say that

---

eighteenth century, but they soon procured horses, abandoned their fields, and took to following the buffalo. Some of the Cheyennes are said to have continued to plant a little corn each year, well on into the nineteenth century.

they first reached the Missouri at the mouth of White
River, in South Dakota, whence they gradually ascended
the Missouri to Heart River, where they lived for a long
time and perhaps reached the height of their prosperity, about two hundred years ago. Here the Mandans
were joined by two Siouan tribes, the Hidatsas and
Crows. But little can be even surmised as to the route
of migration of these peoples. Their traditions say
that they came from the east, from the region of Devil's
Lake, and joined the Mandans on Heart River while
that tribe was still prosperous and strong. According
to their stories they learned all that they knew of agriculture from the Mandans after joining that tribe on
the Missouri. Matthews says: "The Hidatsas claim
to have had no knowledge of corn until they first ate it
from the trenchers of the Mandans. . ."[2] When
they reached the Missouri the Hidatsas and Crows were
one nation, but after living on Heart River for some
time they quarrelled and separated, the Hidatsas remaining with the Mandans, while the Crows moved westward into the plains, abandoning agricultural pursuits
and adopting the life of wandering hunters. They did
not plant corn again for a hundred years.[3]

The Omahas and Ponkas also came to the Upper Mis-

----

[2] *Ethnography and Philology of the Hidatsa Indians*, p. 11.

[3] Agent Frost, in his report for the year 1878, states that corn
was first planted that year among the Crows. The crop was put
in by the agency employes, but a number of Crows helped to care
for the corn and promised to plant patches of their own the following year. They told the agent that long ago the Crows "had
no horses and raised corn." *Report* of the Commissioner of Indian Affairs for the year 1878, p. 86.

By permission of the Montana Agricultural Experiment Station

1. RED LAKE CHIPPEWA   2. MANDAN WHITE FLINT
   FLINT

3. LA POINTE CHIPPEWA   4. FORT TOTTEN SIOUX
   FLINT

souri from the east. There is evidence that these two
Siouan peoples reached the Missouri before the year
1700, but perhaps not earlier than 1675. Their arrival
was therefore later than that of the Mandans, and perhaps than that of the Hidatsas and Crows. The traditions of these tribes tell of their migration northward
through the State of Iowa to the vicinity of the pipestone quarry; then west to the Big Sioux River, where
they were attacked by enemies and forced to remove to
the Missouri River, in South Dakota. They next moved
up the Missouri, crossing to the west bank at the mouth
of White River. The soil here was poor, and after remaining for a short time the tribes moved down the Missouri to a point opposite the mouth of James River.[4]
They lived here for many years, then moved down to the
Niobrara, where the Omahas planted some patches of
corn, beans, and pumpkins. At this time the Ponkas
did not cultivate the ground; they were a hunter tribe,
procuring some corn and vegetables from the Omahas in
barter for dried meat.[5] The traditions state that when
they reached the Niobrara the Ponkas numbered three
thousand people and encamped in three large concentric
circles, the Omahas in two circles. The Ponkas remained

[4] This is the Omaha tradition, as given by Henry Fontenelle
in 1884, in *Transactions* of the Nebraska Historical Society, v. i,
p. 78. Other versions do not mention this halt opposite James
River.

[5] The Ponkas appear to have taken up, and then abandoned,
agriculture several times in early days. Thus Merrill was informed by some traders in 1834 that the Ponkas had formerly
cultivated the soil but had given up the practice. *Transactions*
Nebraska Historical Society, v. v, p. 170.

on the Niobrara, while the Omahas moved down to Bow Creek, and later to Omaha Creek, near Homer, Nebraska, in which vicinity they still reside at the present day.

Two more Siouan tribes, the Otoes and Iowas, followed the Omahas and Ponkas to the Missouri in the year 1700. Crossing to the west bank of the river, the Iowas built a village on Iowa Creek, between the Niobrara and Omaha Creek.[6] They later moved farther south and built a village near the present city of Omaha. The Iowas did not remain long in Nebraska, but moved eastward into the state that now bears their name, where they remained for nearly one hundred years. In late years they crossed the Missouri again and were established on a reservation in northeastern Kansas.

The Otoes after crossing the Missouri about the year 1700 are said to have gone down to the Platte at once, where they built a village at the mouth of the Elkhorn.[7] They remained on the lower Platte, building many villages, until after 1840. In 1778 the Otoes were joined by the remnant of a kindred tribe, the Missouris, who had long dwelt on the lower Missouri, near the mouth of Grand River. These people, although few in num-

---

[6] This is the Omaha tradition, given by Henry Fontenelle, 1884, in *Transactions* Nebraska Historical Society, v. i, p. 78. The Iowas are said to have built two or three villages on the Missouri between the Niobrara and the Platte at this period. Bourgmont mentions them in connection with the Otoes in 1724, but does not locate their village. (*Margry*, v. vi, pp. 396 and 410).

[7] This is Henry Fontenelle's version, given in 1884, in *Transactions* Nebraska Historical Society, v. i, p. 78.

By permission of the Montana Agricultural Experiment Station

EARS OF FORT PECK CORN

ber, maintained a separate village for themselves near that of the Otoes until after the year 1850, and accompanied the Otoes on their removal to the Indian Territory in 1882. This once powerful tribe numbered but forty persons in 1885.

The Cheyennes, an Algonquian tribe, migrated from the far north at a very early date and came down to the headwaters of Minnesota River, where they learned to cultivate the soil and built a village near Lac-qui-parle. They later removed to the western branch of the Red River of the North which still bears their name,[8] where they built another village and cultivated the soil until, early in the eighteenth century, they were attacked by enemies armed with guns, who forced them to remove to the Missouri, apparently about the year 1740.[9] They crossed the Missouri, seemingly above the Arikara villages and below those of the Mandans, and built an earth-lodge village of their own above Standing Rock and just below the mouth of Porcupine Creek,[10] and

[8] This stream is still known as the Sheyenne River. The old Sioux name for this river was *Shaien wojubi* (''place where the Cheyennes plant'').

[9] Alexander Henry, perhaps our best authority on this point, states that the Cheyennes were driven from the Red River region about 1735. They do not appear to have gone direct to the Missouri as the Mandans told the elder Verendrye in 1739 that the next tribe below them on the river was called Panana (Arikara). But the younger Verendrye appears to have met the Cheyennes, above the Arikaras and below the Mandans, in 1743, although he speaks of them as Sioux — an error also made by Carver many years later.

[10] The ruins of this village were still to be seen some years ago, but the whole site has now fallen into the river.

here again they planted their corn patches. The Cheyennes now procured horses from the hunter tribes in the plains to the west, and after a time they abandoned their earth-lodges and gradually moved out into the plains, toward the Black Hills, where game of all kinds was very abundant. They were now a hunter tribe, living in skin tepees, but for nearly a century they continued to plant some corn here and there in small patches.[11]

The Arapahoes, another Algonquian people, state that they met the Cheyennes in the plains, north of the Missouri, and moved with them across the Missouri. The Cheyennes deny this story. The Arapahoes also claim that they planted corn while living north of the Missouri, but that some time after they had crossed that river the Arikaras "stole" their corn.[12] They then moved out toward the Black Hills, took to hunting the buffalo, and soon secured horses.

The Sioux or Dakota reached the Missouri about the middle of the eighteenth century, from the Minnesota

---

[11] Perrin du Lac states that the Cheyennes still planted some corn and tobacco in 1802. Elk River informed Mr. Geo. Bird Grinnell that the Cheyennes built two earth-lodge villages. The first one was "above the Standing Rock." From here part of the tribe moved out toward the Black Hills to hunt; the rest moved down and built a village near the mouth of Cheyenne River. Elk River's mother, born 1786, pointed out the site of this village in 1877 and stated the Cheyennes still lived and planted there when she was a girl.

[12] This is the statement made by Little Raven to Lieutenant Clark in 1881. The evidence, however, all points to the conclusion that the Arapahoes gave up agriculture at a much earlier period than did the Cheyennes.

region, where they had lived when the first white men met them. They were a numerous and fierce race of hunters and warriors, the terror of all their neighbors and the oppressors of all the weaker tribes. From the viewpoint of the agricultural tribes they were to be classed with the smallpox, the drought, and the grasshopper, as one of the great plagues of existence. Some bands of the Sioux practiced agriculture in a desultory way at times; but for the most part the western bands depended for their food on the hunt and the theft or purchase of a little corn from their neighbors.

Of the other hunting tribes in the Upper Missouri area little need be said at this point. To the northeast and north of the Missouri were the Assiniboin, certain bands of Chippewas and Plains Crees, and the three tribes of Blackfeet. South and west of the river were the Crows, Cheyennes, and Arapahoes, and, in earlier times, the Kiowas, Prairie-Apaches, and Comanches. The Atsinas, kinsmen of the Arapahoes, sometimes roved north of the Missouri and sometimes south of it.

At the beginning of the eighteen century the sedentary tribes of the Upper Missouri were all numerous and strong peoples. Renaudiere, who is one of our best early authorities, states that the Pawnees in 1723 had one village of 150 lodges on the Platte below the Elkhorn and eight more villages on the Elkhorn.[13] A Frenchman who had been trading on the Missouri for many years informed Colonel Bouquet in 1763 that the Pawnees then had 2,000 fighting men, which would in-

---

[13] Renaudiere's report, August 23, 1723, in *Margry*, v. vi, p. 392.

dicate a total population of about 10,000.[14]   As late as
1820 this tribe had four strong villages.   In 1872, before the final blows fell upon them, the Pawnee census
was as follows: Chaui tribe: men 140, women 254,
children 365, total 759.   Kitkehahki tribe: men 124,
women 208, children 218, total 550.   Skidi tribe: men
154, women 232, children 244, total 630.   Pitahauerat
tribe: men 91, women 182, children 235, total 508.   All
Pawnees: men 509, women 876, children 1,062, total
2,447.[15]   In 1906 the Pawnees numbered 649.

All of our early informants agree in stating that the
Arikaras, like the Pawnees, were formerly a numerous
people.   Trudeau (1796) found the Arikaras living in
two villages, but he lays emphasis on the fact that the
population of these villages was made up of the rem-
nants of many tribes or bands, formerly independent.[16]
"In ancient times the Ricara nation was very large; it
counted thirty-two populous villages, now depopulated
and almost destroyed by the smallpox which broke out
among them three different times.   A few families only,
from each of the villages, escaped; these united and
formed the two villages now here . . . upon the
same land occupied by their ancestors. . . This na-

---

[14] If anything, this estimate is too small.   The trader had perhaps seen only the Pawnee villages on the lower Platte.   As late
as 1838 the Pawnees still had about 10,000 people, and even
after the great cholera attack of 1849 they had 4,500.

[15] Report of the Pawnee agent, 1872.   This is the only detailed census of the Pawnees ever made, as far as known.

[16] Several dialects were spoken among the Arikaras, and there
were other evidences that the people had formerly been divided
into separate tribes.

tion formerly so numerous, and which, according to their reports, could turn out four thousand warriors, is now reduced to about five hundred fighting men, as I have said, and what is more the lack of harmony which exists among the Chiefs has caused the nation to be divided into four parts.''[17]

Lewis and Clark in 1804 found the Arikaras living in three villages at the mouth of Grand River, South Dakota. They state that these people are the remnant of ten powerful tribes, and the population in 1804 is given as 600 warriors, 2,600 people. In 1859 the tribe had only 109 lodges left.[18] In 1904 the Arikaras numbered 380.

We have similar evidence as to the early strength of the Mandan tribe. Verendrye (1738) informs us that there were five large villages on the bank of the Missouri and one smaller one away from the river on the east bank. Other early informants state that there were nine villages. Thus Lewis and Clark were told in 1804 that ''Within the recollection of living witnesses, the Mandans were settled . . . in nine villages . . . about eighty miles below (*that is, near Heart River*) . . . seven on the west and two on the east side of the Missouri. The two . . . wasting before the smallpox and the Sioux united into one village and

[17] Journal of Jean Baptiste Trudeau among the Arikara Indians in 1796, in Missouri Historical Society *Collections*, v. iv, no. 4, p. 28.

[18] Statement of Wm. G. Hollins of the American Fur Company, in the *Report* of the Commissioner of Indian Affairs for 1859, p. 120: ''Rees, 109 lodges; Mandans, 33 lodges; Gros Ventres, 94 lodges.''

moved up the river opposite to the Ricaras. The same causes reduced the seven to five." [19]

In 1804 Lewis and Clark found the Mandans in two villages at the mouth of Knife River, North Dakota, and estimated their population at 1,250. In 1837 their number was given as 1,600, but in that year they were attacked by the smallpox again, and only 150 survived.[20] In 1859 there were thirty-three lodges of Mandans, according to Wm. G. Hollins. By 1871 the Mandans had increased to about 450, but they then began to decrease again, and in 1905 there were only 249 left, but a handful of whom were of pure Mandan descent.

The early strength of these Upper Missouri village-tribes was broken by repeated attacks of the smallpox, after which they were still farther reduced by the constant harrying of the Sioux, Assiniboin, and other enemies. Trudeau states that the Arikara had suffered from smallpox on three occasions prior to 1795, and Lewis and Clark give similar evidence in regard to the Mandans. The first visitation of the smallpox of which we have any record was the great epidemic of 1780 which worked havoc in the Upper Missouri villages and then spread among the tribes of western Canada where, according to the estimates of the fur traders, one-third of the Indian population died of the disease.

There were no Sioux living near the Upper Missouri as late as 1743, but they appear to have reached the

[19] *Original Journals of Lewis and Clark*, v. i, entry of Oct. 22, 1804.

[20] The number of Mandans who survived this terrible epidemic is variously given as 31, 125, 145, the latter figure evidently the most reliable one.

Missouri River in considerable numbers some time between the years 1750 and 1770. About 1775 an Oglala Sioux named Standing Bull visited the Black Hills; but this was only an incidental journey, and we have sufficient evidence that the Sioux did not cross the Missouri, to hunt and live, in any force until about the year 1800.

The Cheyennes state that the first Sioux who visited them in the Black Hills region were poor people. They came out first (about 1760) in small parties and family groups, to beg meat and horses from the Cheyennes; but they soon came in larger bands and about 1785 began attacking the Black Hills tribes. Some thirty years later, having in the meantime driven out all of these tribes, with the exception of the Cheyennes, they occupied the Black Hills country and claimed it as their own.

We have accounts of the Sioux methods of harrassing the agricultural tribes, written by Perrin du Lac in 1802, Lewis and Clark, and many other early travelers. A typical example of a "friendly" visit by the Sioux to one of these Indian villages occurred in the fall of 1853. A large camp of about 2,500 Sioux came to the Arikara village to trade meat and robes for corn, beans, and dried pumpkins. Having finished trading, the Sioux stole everything they could lay their hands on and then set out for their hunting-grounds, firing the prairies as they went on all sides of the Arikara village — "an act of dastardly malignancy, as it deprived the Arickarees of the means of support for their horses . . ." and also prevented any game from coming near the village.[21] Some of the Sioux bands were us-

21 *Pacific Railroad Surveys*, v. i, p. 265.

ually on this sort of "friendly" terms with the village Indians. The rest of the Sioux were openly hostile, and for a hundred years, from 1775 to 1875, the sedentary tribes, from the Pawnees and Otoes in the south to the Mandans, Hidatsas, and Arikaras in the north, were constantly under the pressure of Sioux hostility. The Assiniboins and other tribes occasionally attacked the villagers; but the Sioux danger was ever-present.

2. The earth-lodge village

The agricultural tribes of the Upper Missouri were also the only persistently sedentary, village-dwelling peoples of the region. The permanent villages of these tribes, except for minor differences, were very much alike. In early times they were usually located on some point of the higher benches or bluffs along the river, whence a clear view might be had in all directions, and where suitable soil was to be found in the bottom land at the foot of the bluff, on the wide bench in rear of the village, or in the bends of nearby creek valleys. Among the Pawnees, Otoes, and Omahas little appears to have been provided by way of defensive works, at least in later days; [22] but farther north on the Missouri the villages were always surrounded by heavy palisades of large logs and earth, when built on the lower bench, and when built on the bluffs one or more sides of the village were protected by the steeply scarped edges of the bluff and only the sides open to attack were surrounded by

[22] Omaha tradition states that they and the Ponkas lived in a strong "fort" on the Big Sioux River before they removed to the Missouri.

palisades and earthworks. Either inside or outside the palisade was a ditch, the location of which varied in the different villages, if we are to believe our early informants. The villages seem to have been much more strongly fortified in early times than they were after the Indians procured firearms. Verendrye, who visited the Mandans in 1738, draws a very imposing picture of the strong wall about their "fort," with the ditch outside some twenty feet across and fifteen feet deep. He states that access was possible only by means of a bridge across the ditch and that there were strong bastions and earth ramparts along the walls. An examination of some of the earlier Mandan village sites seems to corroborate, at least to a certain extent, Verendrye's account of earthwork fortification; while farther south in the old Pawnee country along the lower Platte, we have clear evidence that the villages of the later stone age were much more strongly defended than those of a later time.[23]

Trudeau (1796) gives the following description of the Arikara fortifications: "The Ricaras have fortified their village by placing palisades five feet high which they have reinforced with earth. The fort is constructed in the following manner: All around their village they drive into the ground heavy forked stakes, standing from four to five feet high and from fifteen to twenty feet apart. Upon these are placed cross-pieces as thick

[23] The various reports of the Nebraska State Archaeologist show that most of the stone age village sites in this state were either located on high ground in very strong defensive positions or were surrounded by earthworks and ditches, the remains of which are often still clearly visible.

as one's thigh; next they place poles of willow or cottonwood, as thick as one's leg, resting on the cross-pieces
and very close together. Against these poles which are
five feet high they pile fascines of brush which they
cover with an embankment of earth two feet thick; in
this way, the heighth of the poles would prevent the
scaling of the fort by the enemy, while the well-packed
earth protects those within from their balls and arrows."[24]

Let us view then a typical earth-lodge village, situated
on a high bench of the Missouri; in the rear a level plain
leading back to the hills two or three miles away; in
front the river, washing the bank below the village.
On the plain close to the village, if it be Mandan or
Hidatsa, stand a number of fantastic scaffolds on which
the dead, wrapped in cloths or fine robes, are deposited.
Near the cemetery also are to be seen several shrines
consisting of tall poles carrying effigies of some sort and
surrounded by a circle of human skulls. On another
side, close to the village wall, is the smooth and hard
beaten course for their "billiard" game, as the French
called it. Off toward the hills are many horses, under
the watchful eyes of some of the youths who act as herd-
ers. Near the village in several directions the plain
seems to be one large corn field, though in reality broken
up into many small patches. Let us refer to Henry's
account of his visit to the Mandans and Hidatsas, in
which he describes the country from the Amahami[25]

[24] Trudeau, Missouri Historical Society *Collections*, v. iv, no. 1,
p. 23.

[25] Amahami: a small Siouan tribe very closely related in lan-

village to the lower Hidatsa town: "We proceeded on a delightful hard, dry road. The soil being a mixture of sand and clay and rain being infrequent, the heat of the sun makes the road as hard as pavement. Upon each side were pleasant cultivated spots, some of which stretched up the rising ground on our left, whilst on our right they ran nearly to the Missouri River. In these fields were many women and children at work, who all appeared industrious. Upon the road were passing and repassing every moment natives, afoot and on horseback, curious to examine and stare at us. Many horses were feeding in every direction beyond the plantation. The whole view was agreeable and had more the appearance of a country inhabited by a civilized nation than by a set of savages." [26]

Now let us approach the village itself. From a distance some of the early travelers have likened it to a cluster of mole hills. As we approach however the mole hills become larger, we pass through the gate in the palisade and find ourselves among these great, earth-covered dwellings. The lodges are crowded in without apparent order, leaving frequently hardly room to pass between. The ground is packed hard from the beat of many feet, and is mostly clear and clean, as the village, as well as the lodges, is swept out at short intervals. Before each lodge is an open space sufficiently large to accommodate the two-story scaffold upon which meat

guage to the Hidatsas. They maintained a separate village of their own near the Hidatsa and Mandan villages down to the time of the great smallpox epidemic of 1837.

[26] *Journal of Alexander Henry*, Coues edition, p. 344.

and corn are dried and firewood piled in summer. In this season too we must watch where we go, as many of the storage caches which are scattered promiscuously throughout the village are empty and open pitfalls for the feet of the unwary. Henry says: "So numerous about the village are these pits that it is really dangerous for a stranger to stir out after dark." [27]

Near the center of the village is an open space some 200 to 400 feet across, where the dances and various ceremonies are held. In the Arikara village there will be one lodge in this open space, larger than most of the others, and with a cedar post and a large stone before it. This is the holy house or medicine lodge. In the Mandan village there will be no lodge in the open area; in the center however will be a round enclosure of cedar planks, and this is the "Minaki" — the center of all their religious ceremonies. The doors of the lodges surrounding this open space in the Mandan village all face inward toward the Minaki, and it will be noticed that one of these lodges has a queer, flattened front, and before it stand two high poles to the tops of which are fastened grotesque effigies of some of their gods. This is the Mandan sacred lodge. In the Hidatsa village is no open space at all, the houses being crowded in haphazard throughout the whole area.

Having examined the general appearance of the village, let us next look closer at the individual lodges. They are all circular in form with dome shaped roofs and covered entrance ways six to fourteen feet long before the doors. The average house will be perhaps thir-

27 *Henry*, p. 360.

ty to forty feet across and fifteen to twenty feet high, though many will be twice that size and a few even larger. Lodges ninety and even one hundred feet in diameter have been mentioned. Both the lodge and the passage way are thatched and the thatch is covered to a considerable depth with closely packed earth. The roof is then finished either by being smoothly plastered with clay or covered with sods cut and laid like shingles. In the latter case "the grass of the sod continued to grow, and wild flowers brightened the walls and roof of the dwelling. The blackened circle around the central opening in the roof, produced by the heat and smoke, was the only suggestion that the verdant mound was a human abode." [28]

At the top of the dome shaped roof is an opening two or three feet across, to permit the escape of the smoke and light the interior of the lodge. On the roof beside this opening is laid an old bull-boat or a covered wicker frame, to be placed over the smoke hole in bad weather. A rude ladder, usually chopped out of a log, leans against the house, giving easy access to the roof, which is often surrounded by a wooden railing. On the roof are scattered various household articles and probably one or more old buffalo skulls; and here often groups of men collect, sitting or reclining on the sloping roof, to chat and overlook the village. When there was no great hunt afoot or no warlike demonstration requiring their attention, Henry says: "The young men . . . pass the day on the tops of the huts, sleeping in the sun or strolling from hut to hut, eating corn and smoking

[28] *Bulletin* 30, Bureau of American Ethnology, v. i, p. 410.

Missouri tobacco."[29]   Here on the roofs the whole population congregated at times, to watch the return of a war party or observe the approach of a body of strangers.

When we enter one of these lodges and our eyes become accustomed to the dim light, we notice that the floor of the lodge is a foot or more below the level of the ground outside.[30]   In the center of the floor, under the smoke hole, is a shallow depression some five or six feet across and curbed with stone.   This is the fireplace, and about it are set earthen pots of various sizes, wooden bowls, horn spoons and other articles of a culinary nature.   A portion of the lodge near the door is divided off either by a curtain of skins or a plank wall; behind this firewood is piled and it serves also to prevent the draft from the door reaching those seated about the fire. Another space is sometimes partitioned off for the use of the horses.   Maximilian says: "Inside the winter huts is a particular compartment where the horses are put in the evening and fed with maize."[31]   Between the windbreak near the door and the fire is the master's seat — a willow mat arranged with an elevated back, resembling a divan.   Around the wall are the beds, divided off from the rest of the lodge by curtains of skins or reed mats.   Each bed is a separate, box-like structure, entirely closed in except for one small opening; the bed is raised somewhat above the floor and is filled with robes.

---

[29] *Henry*, p. 327.

[30] Some of the tribes, unlike the Mandans, did not excavate the floor of the lodge, but left it on the same level as the ground outside.

[31] *Maximilian's Travels*, p. 272.

The back part of the lodge, opposite the entrance, is the guests' corner; here also sacred objects are kept — the master's medicine-bag, his war paraphernalia, arms and shield, and often a painted buffalo skull. Between this corner and the fire a large wooden mortar is fixed firmly in the ground, and in the mortar or beside it rests a heavy pestle, also made of wood. Le Raye tells us that at the time of his visit to the Mandans (1801-1802) the walls of the lodges were elaborately hung with various kinds of beautiful furs, all well dressed. Maize and dried meat were often piled up here and there on the floor, while many articles of household use were hung upon the large center posts. From these four posts also, among some of the tribes, "the shields and weapons of the men were suspended . . . giving color to the interior of the dwelling, which was always picturesque, whether seen at night, when the fire leaped up and glinted on the polished blackened roof, and when at times the lodge was filled with men and women in their gala dress at some social meeting or religious ceremony, or during the day when the shaft of sunlight fell through the central opening over the fireplace, bringing into relief some bit of aboriginal life and leaving the rest of the lodge in deep shadow." [32]

Among many of the tribes "ceremonies attended the erection of the earth-lodge from the marking of the circle to the putting on of the sods. Both men and women took part in these rites and shared in the labor of building. To cut, haul, and set the heavy posts and beams was the men's task; the binding, thatching, and

[32] *Bulletin 30*, Bureau of American Ethnology, v. i, p. 411.

sodding that of the women. . . Few, if any, large and well-built earth-lodges exist at the present day. Even with care a lodge could be made to last only a generation or two.'' [33]

The Mandans, Hidatsas, and Arikaras usually abandoned their permanent villages during three or four months in the winter and built similar earth-lodge villages [34] in the timber, along the Missouri River bottoms, where there was more shelter and fuel was plentiful, but as soon as the weather became more favorable for the movement of hostile war parties, or a thaw presaged the break-up of the ice in the river with possible floods, the people at once returned to their permanent quarters. Farther south, in Nebraska, the Ponkas, Omahas, Pawnees, and Otoes spent much less time in their permanent villages, not being as closely pressed by the Sioux or other enemies as were the tribes higher up the Missouri. Indeed, these tribes in Nebraska followed the buffalo for a great part of the year, living in skin tepees, returning to their earth-lodge villages only for a brief period at planting-time and harvest.

The Mandans, Hidatsas, and Arikaras on returning to their villages in early spring occupied themselves before it was time to plant the fields in dressing the pelts they had obtained during the winter and in laying up a supply of firewood for the summer. ''When the ice broke up in the spring the Indians leaped on the cakes, at-

[33] *Bulletin 30*, Bureau of American Ethnology, v. i, p. 411.

[34] These winter camps consisted of small, rude earth-lodges, requiring only a short time to build. The people seem to have camped in small bands in winter, in sheltered and timbered bottoms.

tached cords to the trees that were whirling down the rapid current, and hauled them ashore. Men, women, and the older children, engaged in this exciting work, and although they sometimes fell and were swept downstream, their dexterity and courage generally prevented serious accidents." [35]

3. Agriculture

The village Indians of the Upper Missouri, like most of the other sedentary tribes within the present limits of the United States, in early days depended only partially on agriculture as a means of subsistence, gaining at least fifty per cent of their food supply from the hunt. Some of the tribes like the Kitkehahki Pawnees and Ponkas, although living in fixed villages a part of each year, subsisted almost wholly on the hunt. Pike, writing in 1806, tells us that the Kitkehahki were "in point of cultivation . . . about equal to the Osages, raising a sufficiency of corn and pumpkins to afford a little thickening to their soup during the year." [36] We have similar statements from other travelers with reference to the Osages, Kansa, Otoes, and Ponkas. On the other hand, it is stated that the Pawnees (particularly the Skidi), the Arikaras, and Mandans in early times cultivated the soil extensively and usually had an abundant supply of corn and vegetables en caché during the

[35] *Bulletin 30*, Bureau of Ethnology, v. i, p. 85.

[36] *Pike*, Coues edition, v. ii, p. 533. This passage appears to be the source of Gallatin's statement that the Pawnees did not raise enough corn "to whiten their broth," an assertion to which Dunbar took strong exception. Pike's view was certainly correct at the time, as far as the Kitkehahki tribe was concerned.

greater part of the year. Dunbar informs us that the
Pawnees, although they usually had a large quantity on
hand, never traded or sold their corn. On the Upper
Missouri the trade in corn and vegetables was almost
entirely in the hands of the Mandans and Arikaras who,
from the earliest times, had carried on an extensive
trade in these articles with the hunter tribes of the
Plains. In later times they supplied the white fur traders with large quantities of corn. The Omahas appear
to have carried on a small trade in corn with the fur
traders at certain periods in their history.[37]

Most of the early explorers and travelers who journey-
ed through the Upper Missouri country during the
eighteenth and nineteenth centuries have mentioned in
passing the agriculture of the tribes of this region.

Coronado (1541) did not quite reach the Upper Missouri, his farthest north being the Kansas River or the
Big Blue, where he found the Quivira Indians — evidently the Wichitas, close kindred of the Pawnees. Castañeda says of these people: "In some villages there
are as many as two hundred houses; they have corn and
beans and melons; they do not have cotton nor fowls,
nor do they make bread which is cooked, except under
the ashes." [38]

Verendrye, 1738, says of the Mandans, whom he visited in December of that year: "Their fort is full of
caves, in which are stored such articles as grain, food,
fat, dressed robes, bear skins. . . They brought me

---

[37] The trade, however, was of small extent and not at all regular.

[39] Castañeda, in *14th Report*, Bureau of American Ethnology,
p. 577.

every day more than twenty dishes of wheat,[39] beans
and pumpkins, all cooked."[40] "We noticed that in the
plain there were several small forts, of forty or fifty
huts, built like the large ones, but no one was there at
the time.[41] They made us understand that they came
inside for the summer to work their fields and that there
was a large reserve of grain in their cellars."[42] "Wheat
flour pounded for the journey was brought, much more
than was necessary. I thanked them, giving them some
needles which they greatly value. They would have
loaded a hundred men for the journey; in a short time
all hastened to bring me some."[43] Of the Pananas
(Arikaras) and Pananis (Pawnees) Verendrye says:
"In summer they grow wheat and tobacco on the lower
part of the river."[44]

Henry (1805): "The Mandanes and Saulteurs

[39] Verendrye certainly means maize, but he always speaks of
it as *Blée* "wheat" and as *grain*.

[40] Verendrye Journal, in Canadian Archives *Reports*, 1889, p.
21.

[41] This passage in Verendrye's journal has been much puzzled
over. From recent talks with the Indians, I am inclined to think
that Verendrye is here referring to summer-villages near small
tracts of good farming grounds, and that these temporary summer-villages were made necessary because of the lack of sufficient good soil for all of the people in the neighborhood of the
main villages. The Indians inform me that the Mandan village
on the site of the present town of Mandan, N. D., was called
Scattered Village and that it was used as a summer-village by
the people of Big Village, who went there each year to plant in
the fertile Heart River bottom lands. — George F. Will.

[42] "Beaucoup de grain dans les caves en reserve," p. 22.

[43] Canadian Archives *Report*, 1889, p. 25.

[44] Same, p. 19.

(*Souliers*) are a stationary people who never leave their villages except to go hunting or on a war excursion. They are much more agricultural than their neighbors, the Big Bellies (*Hidatsas*), raising an immense quantity of corn, beans, squashes, tobacco" (p. 338).

LaRaye (1801-1802): "These people (*the Arikaras*) are much more cleanly in their persons, dress, and food than the Sioux. These Indians raise corn, beans, melons, pumpkins, and tobacco. Their tobacco differs from that raised by white people. It has a smaller stalk that grows about eighteen inches high, with long narrow leaves, that is used only for smoking" (p. 161). Of the Mandans LaRaye says: "These people keep their lodges and buildings in a state of great neatness. They cultivate the same kind of produce with the Rus (*Arikara*)" (p. 167).

Lewis and Clark (1804): "They (*Arikaras*) cultivate maize or Indian corn, beans, pumpkins, watermelons, squashes and a species of tobacco peculiar to themselves." "The Indians (*Mandans*) brought corn, beans, and squashes which they readily gave for getting their axes and kettles mended."

Bradbury (1811): "I have not seen, even in the United States, any crop of Indian corn in finer order or better managed, than the corn about these three (*Arikara*) villages. They also cultivate squashes, beans, and the small species of tobacco (nicotiana rustica)"[45] (p. 175).

Boller: "They (*Arikaras*) cultivate large fields of

---

[45] An error of Bradbury's — the Arikaras raised nicotiana quadrivalvis.

corn and also pumpkins and squashes, which agreeably vary their diet of buffalo meat'' (p. 33).

Hayden (1855): "The Mandans at this time number about 35 or 40 huts, perhaps near 300 souls, and raise corn, squashes, beans, etc., the same as the Minitarees and Arikaras'' (p. 434). Of the Minitarees or Hidatsas he says: "Small patches of corn, beans, squashes, pumpkins, and a few other vegetables, have been cultivated by them from the earliest period known, even to the present'' (p. 240).

Maximilian (1833): "The Mandans and Manitaries cultivate very fine maize without ever manuring the ground, but their fields are on the low banks of the river . . . where the soil is particularly fruitful . . . They have extremely fine maize of different species'' (p. 241). "The Indians residing in permanent villages have the advantage of the roving, hunting tribes in that they not only hunt but derive their chief subsistence from their plantations which afford them a degree of security against distress. The distress can never be so great among the Missouri Indians as in the tribes that live farther north. The plants which they cultivate are maize, beans, French beans, gourds, sunflowers, and tobacco, of which I brought home some seeds which have flowered in several botanic gardens'' (p. 274).

De Smet, writing in 1867 of the Mandans, Hidatsas, and Arikaras, says: "They cultivate a large field (1,200 acres), raising corn, potatoes, melons, and beans, with no tools but sharpened sticks, with a few spades and mattocks'' (p. 885). Writing of the Arikaras in 1855,

Hayden tells us: "The land is wrought entirely with hoes by the women and the vegetables raised are Indian corn, pumpkins, and squashes of several kinds" (p. 352).

The agent for the Mandans, Hidatsas, and Arikaras stated in his report for 1878 that this had been the best season in the history of the agency. The Indians planted 800 acres. "More than half of this they have prepared with hoes. This has been as nicely planted and as cleanly kept as any farms in Minnesota. I estimate (August 24) that they will raise 15,000 bushels of corn and 5,000 bushels of potatoes, besides a large amount of squashes, beans, turnips, onions, etc." [46]

Of the agriculture of the tribes farther south — the Ponkas, Omahas, Otoes, and Pawnees — we have few early accounts. The description of Omaha agriculture contained in the narrative of Long's expedition (1819-1820) and Dunbar's account of Pawnee agriculture will be quoted farther on.

Chittenden, speaking of all of the agricultural tribes of the Upper Missouri, states: "In regard to subsistence, permanent residence in fixed localities made possible the development of a crude agriculture and around all these villages there were fields of corn" (p. 844). Of the Mandans he says: "These Indians were an agricultural tribe if that term can be appropriately applied to the crude method by which the natives cultivated the soil. The principal produce was maize, but they also raised certain vegetables introduced by the whites."

Chittenden gives too little credit to the Indians for their agricultural efforts. It is true the acreage planted was small. Dunbar gives the size of the family patches

[46] *Report*, Commissioner of Indian Affairs, 1878, p. 32.

among the Pawnees as from one to three acres, the average being closer to one than to three acres; and the acreage of all of the other tribes on the Upper Missouri was in early days about the same as among the Pawnees — sometimes a little less and sometimes a little more. An acre may seem a very small piece of ground to the American farmer of today, but to the Indian woman whose only implements were the digging-stick, the wood or antler rake, and the bone or iron hoe, an acre had a different look. Chittenden also gives a false impression that among these tribes corn was the only native crop. It is true that some vegetables: notably potatoes, onions, and turnips were introduced by the whites; but from De Soto and Coronado down to Long and Maximilian all of our early authorities agree that the tribes had, besides corn, tobacco, and many varieties of squashes, pumpkins, gourds, melons, and beans. They also cultivated the sunflower for its edible seeds. Potatoes were first introduced among the Mandans by James Kipp, about 1832, according to Maximilian. This crop became a favorite one among all of the Upper Missouri tribes, and after 1860 the Omahas and some other tribes grew more potatoes than corn.

What Maximilian calls French beans were perhaps the wild ground beans of the Missouri bottoms, which the Indians used very extensively but did not cultivate.

Hayden's statement as to the length of time during which the Hidatsas had cultivated the soil does not agree with other accounts or with the Hidatsa traditions, which say that they learned agriculture from the Mandans.[47]

[47] Lieutenant Clark and some others who have quoted the Hidatsa origin tradition have stated that the people cultivated

Nearly all of the Upper Missouri tribes have lost the seed of the varieties of melons grown by their forefathers, but there is evidence that all or nearly all of these tribes formerly cultivated melons. LaRaye, 1801-1802, speaks of the Arikara melons, and Lewis and Clark mention the watermelons grown by this tribe. There does not appear to be any record of melons among the Mandans and Hidatsas, but the Omahas grew them, as did also the Pawnees, who are said to still preserve the seed, while seed of one of these old varieties has recently been secured from the Ponkas. The Wichitas still grow these old-time melons. The Kansa tribe grew both muskmelons and watermelons as late as 1819,[48] while Lieutenant Wilkinson in 1806 ate watermelons at the Little Osage village and describes them as round, "the size of a 24-pound shot," and finely flavored.[49]

To sum up, it is evident that all of the sedentary tribes of the Upper Missouri area grew corn, beans, and squashes; that most of them grew pumpkins, melons, gourds, and tobacco, and some of them sunflowers. They were reasonably good farmers, most of the tribes growing enough corn, beans, and squashes to supply all of their needs, and two or three tribes laying up large reserves of corn and vegetables for purposes of trade. The women did all or most of the work, digging the ground,

corn at the time when they came out of the ground near Devil's Lake. Modern versions of the tradition do not include this statement however, but say that at that time the people cultivated ground beans and wild potatoes, two crops that were not really cultivated at all but merely gathered.

[48] *Long's Expedition*, Thwaites edition, v. i, p. 191.
[49] *Pike's Expeditions*, v. ii, p. 541.

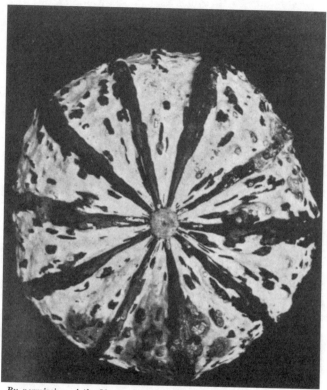

By permission of the Montana Agricultural Experiment Station

MANDAN SQUASH

planting with care, and usually hoeing all of the patches twice before the tribe set out on its summer hunt. The plants cultivated and the general cultural methods corresponded closely to those of the tribes farther east, a good account of which is given in Arthur C. Parker's recent paper on Iroquois agriculture.[50]

The corn raised by the Indians of the Missouri Valley varies from the Pawnee corn, some six to ten feet high, to the corn obtained in late years by the Assiniboins of Montana from the Mandans and by the Assiniboins of Canada from the Sisseton Sioux, which has been acclimated by them and which seldom attains a height of over two feet. From the Pawnee corn in the south we get a regular gradation as we ascend the river, the size of the plants and ears decreasing and the length of season required to mature the crop diminishing.

The varieties of corn grown by the Missouri River tribes will be described at length farther on. A brief survey of the species may be given here. The dent corn to which white men have largely devoted their attention in breeding, was not raised by the Indians of the Upper Missouri; and, with the exception of the Omahas and Pawnees, popcorn seems to have been unknown among these tribes, until introduced by the whites. They raised only three species: the flour, the flint, and the sweet corns.

The flint corn is usually eight rowed, occasionally ten or twelve rowed; this species is high in protein and the grain is very hard and heavy. The flour corn resem-

---

[50] Iroquois uses of maize and other food plants, by Arthur C. Parker, New York State Museum, *Bulletin 144*, Albany, 1910.

bles the flint in the number of rows and in general appearance, but the grain is much softer and lighter in weight, being largely composed of starch and deficient in the proteins. This species was the most popular with the Indians, as it could be easily crushed or ground and was much softer than the flint when eaten parched. This is the species usually referred to by many writers who have stated that Indian corn was of little value as a stock food. The sweet corn is high in sugar content, and the grains when ripe and dry always have a shrivelled and wrinkled appearance.

All of these three species were represented in the Upper Missouri Valley by many varieties of diverse colors and periods of maturity. Of the corn plant, as raised by the Upper Missouri tribes, we have several good descriptions by early travelers. As has been said, the size of the plants and ears varied greatly. This variation was not a matter of latitude alone, but was sometimes due to special seasonal or soil conditions. It is this adaptability to conditions of climate and soil which accounts for the extreme hardiness of the Indian varieties of corn and for the fact that a failure of the crop was rarely absolute.[51] In a poor year and in poor soil the Mandan varieties will seldom rise higher than two feet, the ears being not over three or four inches long with very few to a hill.[52] In a good season, and es-

---

[51] The only complete crop failures mentioned on the Upper Missouri in early times appear to have been caused by the "grasshoppers" which descended in dense clouds and in a few hours stripped the corn patches bare.

[52] Catlin must have been on the Upper Missouri in a bad year, as he says that the Mandan corn produced ears no longer than a

pecially in the rich bottom soil of the river valley, the same varieties will attain a height of five or six feet and will sucker profusely, producing many ears to a hill, some of which will be seven to eleven inches in length.

Another attribute of this northern Indian corn is the very heavy foliage, making a hill appear almost like a bush. This characteristic is very marked in cross breeds and distinguishes the improved crosses which contain a northern Indian strain. Bradbury, himself a botanist, said of the Hidatsa corn in 1811: ''On our approach some fields of Indian corn lay betwixt us and the villages . . . the corn was now nearly a yard high . . . This is about the full height to which the maize grows in the Upper Missouri, and when this circumstance is connected with the quickness with which it grows and matures, it is a wonderful instance of the power given to some plants to accommodate themselves to climate. . . This plant is certainly the same species of zea that is cultivated within the tropics, where it usually requires four months to ripen and rises to the height of twelve feet. Here ten weeks is sufficient, with a much less degree of heat. Whether or not this property is more peculiar to plants useful to men, and given for wise and benevolent purposes, I will not attempt to determine.'' [53]

Hayden (1855) also gives a very brief account of the northern Missouri River corn: ''The corn is said to be the original kind discovered with the continent and

man's thumb. With good treatment in a favorable year this corn today produced large plump ears, seven to eleven inches long.

[53] *Bradbury*, Thwaites edition, p. 159 (and *footnote*).

is quite different in appearance from that raised in the states. The stalk is from three to six feet in height, seldom more than four or four and a half feet, and the ears grow in clusters near the surface of the ground. One or two ears sometimes grow higher up on the stalk, which appears too slender to support any more grain. The grain is small, hard and covered with a thicker shell [*husk*] than that raised in warmer climates. It does not possess the same nutritive qualities as feed for animals as the larger kind, but is more agreeable to the taste of the Indians. It is raised with so little labor that it seems well calculated for them. An acre usually produces about twenty bushels'' (p. 352). Boller says of the corn of the Mandans, Hidatsas, and Arikaras: ''It is a species of Canada corn, very hardy and of quick growth. It is of all colors; red, black, blue, yellow, purple, white; sometimes a single ear presents a combination of all these hues'' (p. 118). By Canada corn he probably means flint corn, as that name was usually applied to the most popular varieties of that type in the eastern states in his day.

Summing up the physical characteristics of the northern Missouri River corn: It is short, rarely attaining a height of six feet, and often below four. It is very much addicted to suckering or stooling from the root, and its foliage is heavier than the average. The ears are from three to nine or ten inches long, usually eight rowed in all three species (flour, flint, and sweet corn), and they grow very close to the ground, often emerging from the ground on a short stalk or shoot from the main stalk. There are frequently two ears to a stalk and three are not exceptional in a very good season. It is extremely hardy, not only adapting itself to varying amounts of moisture and producing some crop under drought conditions, but resistant also to the unseasonable frosts which are apt to occur in the home region. It will sprout in spring weather that would rot most varieties of corn, and once sprouted it grows very rapidly. Its period from planting to maturity is about sixty days in a favorable year, and rarely are more than seventy days required. We may add here an extract from a recent publication of the Montana Agricultural College Experiment Station: [54]

"Early Flints. These varieties originated in North Dakota and contain 'blood' of the old Mandan Indian corn. The stalks grow about four feet in height and are very leafy. The ears are borne close to the ground and cannot be cut with a corn-binder. These corns will stand more hardship in the way of droughts, poor cultivation, frosts, etc., than any other kind. They have been the highest yielders of grain, as a group, of all of the varieties tested in Montana on dry land. They are the earliest matured, withstanding hail well, and make a crop even in a very dry year. They are recommended to be grown all over the state where the season will permit them to ripen."

The varieties grown by the Pawnees and other tribes in Nebraska were very much like those grown farther north. Thus we find that the Pawnees and Mandans both have the same varieties of white, yellow, and blue

---

[54] *Bulletin 107,* "Corn in Montana," by Alfred Atkinson and M. L. Wilson, Bozeman, Mont., October, 1915.

flour corn today, and they formerly had the same varieties of black, red, and red-striped corn. These southern strains attain a larger growth and produce larger ears than do the northern ones. The Pawnee corn grows eight to ten feet high; the plants are given to suckering and the ears are borne much higher on the stalk than in the case of the Mandan and other northern varieties. None of the early travelers on the Upper Missouri have left descriptions of the corn of the Pawnees, Otoes, Omahas, and Ponkas, and what little we know of the corn of these tribes has been gleaned from the planting of small patches of many varieties during the past two or three years. It may be said, however, that these southern varieties seem as hardy and well adapted to meet the conditions in their home region as are the varieties grown higher up the Missouri.

As was usual with most of the tribes of the United States, the women of the Upper Missouri did most of the village work; not only were they the cooks and housekeepers, but the farmers as well and, as we shall see, very good farmers.

Maximilian states (Mandans): "The building of the huts, manufacture of their arms, hunting, and wars, and part of the labors of the harvest are the occupations of the men. The women . . . lay out the plantations, perform the field labor, etc." (p. 271).

Catlin (Mandans): "This (the gardening) is all done by the women."

Bradbury (Arikaras): "The women as is the custom with the Indians do all the drudgery, and are excellent cultivators" (p. 175).

Matthews (Hidatsas): "Every woman in the village capable of working had her own piece of ground" (p. 11).

Dunbar (Pawnees): The women "dug the ground, planted, hoed, gathered, dried, and stored the corn." [53]

Long (1819) states that the Omaha women did all of the field work.

Most of these early accounts give the impression that the women were drudges who were forced to perform most of the heavy labor, including that of the fields. To those who are acquainted with the Indians it will be easily understood that this conception of the position of Indian women needs to be considerably modified. While there is no question that the women's work was severe, yet there is abundant evidence that the women performed their tasks willingly and took great pride in doing their work well. To those who have seen the Indian woman patiently and solicitously working about her garden it must be evident that she loved her work there and enjoyed it.

As a matter of fact in the Upper Missouri region the spring was longingly awaited as the time to commence work on the gardens which furnished much of the pleasure of the summer season; and the harvest time, though a season of rejoicing, yet was also a time of regret for the pleasant summer passed. The Indian woman was a real gardener. Her methods were not those of the bonanza farmer of the present day, but resembled more closely those of the modern market gardener or greenhouse man. She attended to every little detail, work-

---

[55] *Magazine of American History,* v. ii, p. 264.

ing slowly and carefully and taking the utmost pains. She knew the habits of each of her plants and the habits of each separate variety of all the species cultivated, and she worked with careful regard for these differences.

II—PLANTING AND CULTIVATION

1. Spring work: clearing and planting the ground. 2. Hoeing and weeding. 3. The patches, acreage, and yields

1. Spring work: clearing and planting the ground

On the Upper Missouri the agricultural operations for the year began as soon as the ground was sufficiently thawed in the spring, or even earlier if a new plot was to be cleared for use. In the latter case all of the brush was first cut off, and of the larger trees all except one or two were girdled.[1] Then the brush was piled in heaps and burned. In an old garden the work usually started when the first geese appeared on their way north or when the Missouri River broke up, events which usually occurred almost together. At this time the old weeds and stalks and vines which had not been disposed of in the fall were collected and burned.[2]

Of the Hidatsas Henry says: "In the spring they return from the winter villages to sow their fields, while the men are employed getting driftwood and drowned buffalo from the river" (p. 349).

Boller has considerable to say of the spring work

---

[1] This refers to the tribes that cultivated the Missouri River bottoms, where the land was often timbered. The Pawnees and Otoes lived away from the Missouri in almost timberless country.

[2] Dr. Wilson states that the Hidatsas began planting when the wild gooseberries were in full leaf.

among the Mandans, Hidatsas, and Arikaras: "In the spring, as soon as the frost is out of the ground, the women break up their patches of land, every foot must be turned up and loosened with the hoe — a slow and toilsome operation. . . While the operation of breaking the ground, planting and fencing is going on, wood has also to be carried to the lodges, for those great round, earth covered dwellings of the Minnetarees are very chill during the early, damp spring weather; requiring much fuel for warming as well as cooking" (p. 118).

Dunbar states that the Pawnees remained on their winter hunt in the Plains until the young grass began to appear early in April, and then returned to their villages to clear and plant the patches. On another page he says that the patches were cleared out as soon as the frost was out of the ground.

Dougherty states that the Omahas also returned to their village from the winter hunt in April, to clear and plant their patches.

In April, while they were clearing up the corn patches, the women were also busily engaged in dressing the buffalo robes and peltries collected during the winter. Many of the early travelers state that at this period of the year the men did nothing, sitting around smoking and chatting while the women toiled; but upon a closer examination of the available sources it becomes evident that these statements are misleading. Thus we find that the men of the Mandans, Hidatsas, and Arikaras engaged in hunting while the women cleared the patches, and that they were also occupied in catching driftwood and drowned buffalo among the floating ice cakes in the Missouri. In April the men of the Omahas were busily engaged in hunting, often making trips of seventy or eighty miles in quest of deer and elk; while at the same time they trapped all of the small creeks within many miles of the village for beaver, otter, and muskrat.

After the fields were cleared, the hills were dug up with the digging-stick, the ground being pried up, loosened, and pulverized. These hills were about twelve to eighteen inches in diameter and a good long step apart.[3] The ground between the hills was not usually broken up, but the same ground was planted year after year, and in the older patches the position of the hills was sometimes shifted, until finally the whole area of the garden had been dug over. Sunflowers were usually planted around the edge of the garden before the clearing and digging work were completed. This crop was the first to be put in; the Mandan time for planting is given as when the river breaks up.

After the ground had been cleared and well stirred up, the planting of corn and vegetables commenced. It is interesting to note the names of the different moons among these tribes.

---

[3] Buffalo Bird Woman states that in the old days when clearing a new field the women dug the corn hills with digging-sticks, and after planting the corn they dug between the rows with bone hoes. In later times they performed both these operations with iron hoes. In clearing new ground the women did all of the work, a few of the older men helping them at times. They cut the willows and brush in the fall, and in the spring spread them evenly over the ground and burned them. This was done to make the ground softer and easier to dig.

Maximilian states that the Mandans call May the moon
when maize is sown. Clark (*Indian Sign Language*)
says the ninth moon in the old Cheyenne calendar (evidently May) was called "The corn is planted." Pres-
cott, 1850 (in Schoolcraft, *Indian Tribes*, v. ii), gives a
list of Sioux moons: May, corn is planted; June, corn is
hoed; August, corn is gathered; September, wild rice is
gathered. The Omahas called May the planting moon.
The Pawnees called April the field-cleaning moon; May,
the planting moon; June, the cultivating moon; and
July, "Moon-plant-search-complete, or Burning Moon." [4]

Hayden: "Indians plant about the middle of April
or the beginning of May, according to the mildness or
severity of the spring, and the ears are gathered about
the beginning of August" (p. 352).

Merrill (May 12, 1834): "It is now planting time
(among the Otoes). The men lay upon their couches
or sit upon the ground and smoke their pipes all day
long; while the women go from half a mile to two miles
to plant their corn, often, too, carrying a babe with
them. They are also required to bring their wood and
water, which are half a mile distant." [5] On the following day Merrill visited the patches, which were along
the margins of small creeks near the Platte. In Chief
Ietan's patch three of the chief's wives and some other
women were digging up the soil with heavy hoes.

---

[4] Meaning the moon when the corn is found to be complete
(that is, fit for use) or the (Corn) Roasting Moon? See Dorsey, *Traditions of the Skidi Pawnee*, p. 203, for a full list of
Pawnee moons.

[5] The Diary of Rev. Moses Merrill, in Neb. State Hist. Soc.
Transactions, iv, p. 164.

Of the Omahas Dougherty states: "In the month of May they attend to their horticultural interests, and plant maize, beans, pumpkins, and water-melons, besides which they cultivate no other vegetable." [6]

In a recently published account of the Omahas, by Miss Alice Fletcher and Francis La Flesche,[7] we have a more detailed description of the planting operations of this tribe: "The bottom lands were the planting places; each family selected its plot, and as long as the land was cultivated its occupancy was respected. Corn, beans, squash, and melons were raised in considerable quantities" (p. 95). "Garden patches were located on the borders of streams. Occupancy constituted ownership and as long as a tract was cultivated by a family no one molested the crops or intruded on the ground; but if a garden patch was abandoned for a season then the ground was considered free for anyone to utilize. Men and women worked together on the garden plots,[8] which ranged from half an acre to two or three acres in extent. Occasionally a good worker had even a larger tract under cultivation. These gardens were mounded in a peculiar manner: The earth was heaped into oblong mounds, their tops flat, about 18 by 24 inches, and so arranged as to slant toward the south. The height on the north side was about 18 inches; on the south the plot was level with the surface of the ground. These mounds were 2 or 3 feet apart on all sides. In one

---

[6] Long's Expedition, 1819-1820, Thwaites edition, v. i, p. 289.

[7] "The Omaha Tribe," by Fletcher and La Flesche, *27th Report*, Bureau of American Ethnology.

[8] All of our early informants state that the Omaha men did not assist the women.

mound seven kernels of corn were scattered; in the next mound squash seeds were placed, and so on alternately. If the family had under cultivation a large garden tract the beans were put into mounds by themselves and willow poles were provided for the vines to climb upon; but if ground space was limited the beans were planted with the corn, the stalk serving the same purpose as poles. Squash and corn were not planted together, nor were corn, beans, and squash grown together in the same mound. After the planting the ground was kept free of weeds and when the corn was well sprouted it was hoed with an implement made from the shoulder blade of the elk. The second hoeing took place when the corn was a foot or more high. Up to this time the mounds were carefully weeded by hand and the earth was kept free and loose. After the second hoeing the corn was left to grow and ripen without further cultivation. The mounds containing the squash and those in which the melons were planted were weeded and cared for until the second hoeing of the corn, when they, too, were left, as about this time the tribe started out on the annual buffalo hunt'' (p. 269). ''In clearing the ground for planting, the heavy part of the work was not infrequently done by men as were the cutting and transporting of the large posts needed for building the earth lodge'' (p. 339).

The Rev. Gilbert L. Wilson gives the following account of planting among the Hidatsas: [9] ''Corn plant-

---

[9] Corn in Montana, *Bulletin 107*, Montana Agricultural College Experiment Station, Bozeman, Mont., 1915, p. 43.

I have recently obtained the following additional information

ing began in the first half of May, after sunflower seed had been planted. The field was raked free of debris and the stalks of last year's crop, and the dried piles of debris were burned. The corn was planted in hills, three or four feet apart, seven or eight kernels in a hill. The earth was loosened with a wooden digging stick, or with a hoe. Each corn hill stood exactly where a hill had stood the year before. Beans were commonly planted between the corn hills. . . As soon as weeds appeared after planting, the field was hoed. A second hoeing, and hilling up of the plants followed soon after.

"Rotation of crops and fertilization were not practiced, but when a field began to fail, it was let lie fallow for a couple of years. The value of the ashes left from burning over a newly made field was understood.

"Indians insist that corn culture by hoe is much harder now, owing to the abundance of weeds that have been

from the Hidatsas: Fields were marked out by mounds of earth at the four corners. Sometimes another woman would encroach on a field and the matter would be disputed until the intruder was bought or driven out. Corn was planted six or eight kernels in a hill. Often a second planting of corn was made when the june-berries were ripe, in order to secure late roasting ears. Sweet (sugar) corn was never used for roasting ears, hard or soft yellow corn being considered best for that purpose. A sort of bread was made from pounded fresh green corn, but never from ripe corn. Hard white and hard yellow corn were never parched, and therefore were never braided. Seed ears of these varieties were selected from the completely husked out pile of ears on the drying stage. The Hidatsas did not have either watermelons or pumpkins. Sunflower heads were often eleven inches across. Each family planted only one variety of sunflower, one family planting the black, another the white, etc. — George F. Will.

brought in by white men. One curious custom was their carefully removing all dung of horses from the field in the spring, because 'weeds always came up where the dung lay and we thought that white man's cattle and horses brought the seeds.' ''

Buffalo Bird Woman (Hidatsa) states that the crows pulled up the corn plants when they were an inch or so high and the spotted gophers dug up the sprouted seed. At planting time young girls often watched the fields, to frighten off the crows; scare-crows were also set up, but were of little service, and the women often had to go over the fields and replant where the crows and gophers had been at work.

Dunbar states of the Pawnees: The corn patches were usually on some small stream near the village, but sometimes five or more miles away. The ground was cleared and dug as soon as the frost was out of the soil in spring; the corn was planted in early May and was hoed twice, the second time about the middle of June; and immediately after the second hoeing the tribe set out on its summer hunt. He states that where the corn patches were not protected by the high bank of the stream or by some other natural obstacle, they were rudely fenced with bushes and tree branches woven together.[10]

James visited the Pawnee villages with Long's expedition in 1820.[11] ''The agriculture of the Pawnees is extremely rude. They are supplied with a few hoes by the traders, but many of their labours are accom-

---

[10] Dunbar's account of the Pawnees is printed in *Magazine of American History*, v. iv and v.

[11] *Long's Expedition*, Thwaites, *Early Western Travels*, v. xv.

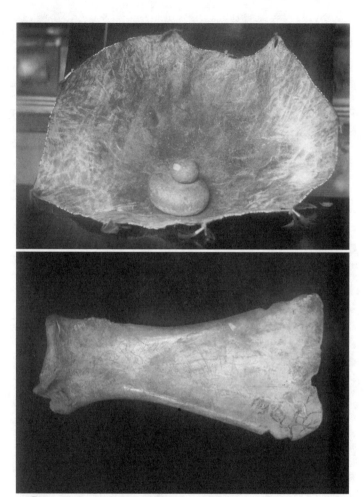

By permission of the Nebraska Historical Society.

Above: RAWHIDE BOWL AND STONE MORTAR
Below: BONE HOE (BUFFALO SHOULDER BLADE)

plished with the rude implements of wood and bone which their own ingenuity supplies. They plant corn and pumpkins in little patches along the sides of deep ravines, and wherever by any accident the grassy turf has been eradicated. Sometimes these little plantations are enclosed by a sort of wicker fence, and in other instances are left entirely open. These last are probably watched by the squaws during the day time, when the horses run at large'' (p. 216). On June 11: ''At a few miles distance from the village, we met a party of eight or ten squaws with hoes and other instruments of agriculture, on their way to the corn plantations. They were accompanied by one young Indian'' (p. 204). ''As day began to dawn on the following morning, numerous parties of squaws, accompanied by their dogs, were seen on their way from the village to the corn patches, scattered at the distance of several miles'' (p. 209). On the 13th: ''As soon as the day dawned we observed the surrounding plain, filled with groups of squaws, with their small children, trooping to their cornfields in every direction. Some, who passed our encampment, lingered a moment to admire our novel appearance; but the air of serious business was manifest in their countenances, and they soon hurried away to their daily labours. Some of the groups of young females were accompanied by a jolly looking young man as a protector'' (p. 217).

Scattered Corn, an elderly Mandan matron, whose father was the last Mandan corn priest, and who is an excellent gardener herself, gives the following information on Mandan planting:

The corn was usually planted in May; that is, the

planting of corn was started then. It was a laborious task, and the work had to be done in the intervals between the usual household tasks. In the larger fields the planting occupied the time from early in May until the time when the roses bloom in June. The rows were measured off about a long step apart, the rows of corn alternating with rows of beans. Seven or eight kernels of corn were planted in each hill, each kernel being carefully placed with the fingers, a small hole being punched in the earth, the grain inserted and the earth carefully patted down. Then the woman moved on to the next hill. If the garden was small, there was an interval after the corn planting before the beans were planted, but in the larger gardens the beans were planted immediately following the corn. The beans were planted in the same manner as corn about six inches apart in the hill. After the beans, the planting of the squash completed the planting work. These were planted when the roses commenced to bloom.

The implements used by the Mandans were the hoe, the digging-stick, and the rake. The digging-stick was a heavy ash pole about three and a half feet long and one and a half inches in diameter. The point was sharpened and hardened in the fire.[12] Of rakes there were two sorts, one from the deer antler, the other made of long willow shoots, one-half to three-fourths of an inch in diameter, cleverly spread and bent at the lower ends

___

[12] Buffalo Bird Woman describes the Hidatsa digging-stick as a stout ash sapling with a slight bend at the root. The end was trimmed to a three-cornered point; this was well greased with bone-butter, wrapped in dry grass, and fired. The slight charring made the point nearly as hard as iron.

to form the teeth, while the upper ends of the long straight whips were bound together with rawhide thongs, to form the handle. The rake and the digging-stick were used only in clearing and preparing the ground for planting; the hoe was the great implement of cultivation. In early times it was usually made of the shoulder-blade of the buffalo or elk, trimmed down and neatly fitted and tied to a wooden handle. Arisa, the Skidi Pawnee priest, stated that the bone blade was attached to the handle with tough gristle taken from the neck of the buffalo.

Bradbury (p. 175): "The only implement of husbandry used by them is the hoe. Of these implements they were so destitute before our arrival, that I saw several squaws hoeing their corn with the blade bone of a buffalo, ingeniously fixed to a stick for that purpose."

Maximilian (p. 276): "At present the women use in the field labors a broad iron hoe with a crooked wooden handle which they obtain from the merchants. Charbonneau recollected the time when they used the shoulder blade of the buffalo."

Henry (p. 343): "I set off early on horseback with part of my people for the upper villages. We passed extensive fields of corn, beans, squashes, and sunflowers. Many women and children were already employed in clearing and hoeing their plantations. These hoes were the shoulder blade of the buffalo to which is fastened a crooked stick for a handle; the soil not being stony, this slight implement answers every purpose."

Matthews (p. 19): "For cleaning the village grounds they had rakes, made of a few oziers tied together, the ends curved and spreading. Their most important agricultural implement was the hoe; before they obtained iron utensils of the white traders, their only hoes were made of the shoulder blades of elk or buffalo attached to wooden handles of suitable length. Maximilian in 1833 considered the bone hoe as a thing of the past only, yet as late as 1867 I saw a great number in use at Ft. Berthold and purchased two or three, one of which was sent to Washington.''

DeSmet (p. 830): ''All their farm implements are a few mattocks and poor spades with crooked or pointed sticks and shoulder-blades of buffalos.'' (P. 885): ''. . . with no tools but sharpened sticks with a few spades and mattocks.'' (P. 974, of the Pawnees): ''. . . using the shoulder-blade of the buffalo as a substitute for the plow and hoe.''

Catlin (v. i, p. 137): ''. . . this is all done by the women who make their hoes of the shoulder blade of the buffalo and the elk, and dig the ground over instead of plowing it, which is consequently done with a vast deal of labor.''

Of the Cheyenne method of planting, Mr. Grinnell states: ''The corn was planted in quincunxes, four grains at the corners of the square, and one in the middle; and the grains were put in the ground with the soft end up. In the occasional plantings which have taken place since 1850, old women and old men have been observed to carry water from the stream to water the hills of corn. It is said that some of the hoes they used were made of stone. Sometimes these stones were chipped to a proper flat shape, but stones naturally of the right form were often found and were lashed to a stick for use. Many hoes were also made of the shoulder bones of the buffalo or elk.''

2. Hoeing and weeding

All of the tribes of the Missouri Valley south of the Arikaras, including the Ponkas, Omahas, Pawnees, Otoes, Missouris, Iowas, Kansa, and Osages, always abandoned their villages and fields late in June or early in July and went out into the Plains on their summer buffalo hunt. They usually gave the corn two hoeings and weeded it thoroughly by hand before going on the hunt; but occasionally when the crops were backward the tribe left after the first hoeing.

On the other hand, the tribes living above the Ponkas: the Arikaras, Mandans, and Hidatsas, usually remained near or in their permanent villages during the entire summer.[13]

Maximilian, p. 368 (Hidatsas): ''At present the Minitaries live constantly in their villages, and do not

[13] Major Marston, writing in 1820, gives a very interesting account of Sac and Fox agricultural methods. These tribes were considered the best farmers of the whole Upper Mississippi region. They planted about 300 acres and grew about 8,000 bushels of corn each year, 1,000 bushels of which they sold to the white traders of Prairie du Chien. The women did most of the work with rude hoes. In spring and summer the men were away, but the women, children, and old men remained at the village, caring for the crops and making bark bags in which to store the crops. In fall most of the crop was buried in these bags, for use in the following spring and summer, but each family took five bushels of corn on the fall hunt. Blair, *Indian Tribes of the Upper Mississippi*, v. ii, p. 152.

roam about as they formerly did, when, like the Pawnees and other nations, they went in pursuit of the herds of buffalo, as soon as their fields were sown, returned in the autumn for the harvest, after which they again went in the prairie.''

From all accounts it appears that the Mandans were the most faithful cultivators of any of the Upper Missouri tribes. They were the most sedentary, and not only is there no record of the whole tribe ever abandoning the villages and fields to go on summer hunts, but it was a very uncommon thing for even a large party to forsake their crops for any considerable time. The Arikaras were also excellent cultivators on occasion, but they frequently took a season off to go on a long summer hunt, and they very often abandoned their old villages to seek a new location.

In the largest gardens of the Mandans the hoeing started immediately after the planting was finished in June and was kept up continuously until late in the summer. This usually gave time to hoe the whole garden through twice. The hoeing was mixed in with the other labors of the day. The women usually got up with the sun, 3 or 4 o'clock A.M., and went immediately to the gardens.

Henry, p. 327: ''The young men employ the night in addressing love songs to their mistresses, who either come out of the huts immediately, or wait till daybreak, when they repair to the corn fields, and are soon followed by the young men.''

They usually worked in the gardens until the sun was well up and its heat began to be felt, when they returned

to the lodges and went about their regular household
duties. Later, any spare time of sufficient length was
also devoted to the hoeing, especially toward the close
of the afternoon. Let it not be thought that this work
was mere drudgery. Every woman had the company
of some of the young girls, and the gardens were close
enough together to permit of friendly intercourse. The
women usually sang as they worked, and there are preserved great numbers of field-songs which were sung only
in the gardens. Some of the young men were sure to
be attracted by the presence of the maidens and usually
hung about while the women worked. In unfavorable
locations there was of course danger from prowling war
parties, and it was often considered necessary to have
an armed guard in the vicinity. This however seemed
to detract in no way from the pleasure of the girls.''

LaRaye's *Journal*, p. 238: ''The village (Hidatsa)
is often visited by their most terrible enemy, that is the
Sioux, who generally succeed to kill some of the inhabitants, especially the woman who is working in her
corn field, for every family has a small patch under
cultivation. The work is done by the squaws.''

Henry speaks of the danger from hostile war parties
and the method of guarding against it (p. 324): ''We
soon met a Mandan, well armed, with his gun, etc., he
accompanied a party of women hoeing corn, and served
as their guard. . . We saw many women and children at work in the corn field.'' And p. 403: ''Even
when they go out to hoe their corn, young men well
armed, keep on the rounds at short distances from the
women to prevent surprise from an enemy. This is a

necessary precaution, as they have frequently been attacked while working in their fields.''

Toward the latter part of the summer, when the ears were beginning to form, a garden was very seldom unoccupied during the day-time. The blackbirds and crows were very numerous about the villages and some one had to be always near to drive them away. For the convenience of the watchers there was usually a brush shelter of some sort, or a scaffold built under the shade of one of the large trees, one or two of which, as we have said, were always left standing when the land was first cleared. Here some of the young girls were stationed, singing, gossiping, and working at some sort of sewing; or perhaps the mother and her daughters brought out their lunch and spent the day there. If a woman was childless it was permissable for her to be accompanied to the field by her husband, and they had their lunch and spent the day together in the garden.

Of the tribes farther south, in Nebraska, we have less detailed information; but Dunbar mentions the same kind of platforms in the Pawnee corn patches, upon which watchers were stationed to frighten away the birds.

We have numerous accounts of the hoeing and descriptions of the bands of women working in the fields.

Henry, p. 345 (between the first and second Hidatsa villages): ''Here the road is again very pleasant, running through an open, level country, with corn fields in sight, in which were numbers of people at work.''

Henry, p. 359 (Mandans): ''The morning was calm and serene; the natives were passing continually to and

PONKA CORN, AT OMAHA, NEB.
(Height, 72 inches)

PAWNEE CORN, AT OMAHA, NEB.
(Height, 7 feet)

fro between the villages; others again were at work in their fields; and great numbers of horses dispersed in every direction served to enliven the scene." And p. 344: "In these fields were many women and children at work, who all appeared industrious."

Boller, p. 48: "Day after day until it was gathered in, the corn must be regularly hoed, more to counteract the effect of drought than to keep down weeds, for in these dry and elevated plains rain seldom falls after the spring is passed. All these duties devolve upon the women."

Chittenden, p. 807: ". . . The corn was hoed two or three times during the season, pains being taken to bank the earth against the hills for the better retention of moisture."

DeSmet, p. 250 (Arikaras): "I was surprised to find around their dwellings fair fields of maize, cultivated with the greatest care."

Of the return to old fields after a lapse of time Boller gives us an instance (p. 324): "The Mandans began to agitate the question of returning to their old village close by the Rees to plant corn in the same field as they tilled years ago when their nation was strong and powerful." Matthews will be quoted on this subject farther on.

3. *The patches, acreage, and yields*

Now that the planting and cultivation of the crops have been described, a more detailed view of the patches may be taken.

Of the patches of the Mandans, Hidatsas, and Arikaras, around Fort Berthold, Matthews gives a very good description:

"From the base of the prairie terrace, described in Sec. 2, the bottom lands of the Missouri extend to the east and to the west up and down the river. In the neighborhood of the village they are covered partly with forest trees, willows, and low brush but chiefly with the little fields or gardens of these tribes.

"Five years ago all the land cultivated around the village consisted of little patches, irregular in form and of various sizes, which were cleared out among the willows. The patches were sometimes separated from one another by trifling willow fences, but the boundaries were more commonly made by leaving the weeds and willows uncut, or small strips of ground uncultivated between the fields. Every woman in the village capable of working had her own piece of ground which she cultivated with the hoe, but some of the more enterprising paid the traders in buffalo robes to plow their land. Their system of tillage was rude, they knew nothing of the value of manuring their soil, changing the seed or alternating the crops. Perhaps they had little need of such knowledge, for when the soil was worn out they abandoned it, and there was no stint of land in the wilderness. Sometimes after a few years of rest they would resume the cultivation of an old field that was quite near the village, for proximity lent some value to the land; but they had no regular system of fallowing. They often planted a dozen grains of corn to the hill and did not hoe very thoroughly. Within the last few years there has been an improvement in their farming; the bottom to the west of the village is still divided up and cultivated in the old way, but the bottom to the east and a part of the upland have been broken up by the Indian Agency, fenced and converted into a large field. A portion of this field is cultivated (chiefly by hired Indians) for the benefit of the Agency, and the rest has been divided into small tracts, each to be cultivated by a separate family for its own benefit. Potatoes, turnips, and other vegetables have been introduced. The men apply themselves willingly to the labors of the field'' (p. 11).

As a gauge of measurement the Mandan Indians had the nupka or Indian acre. This was a somewhat indefinite measurement, consisting as nearly as could be discovered of seven rows of corn with rows of beans between each two rows of corn, and with no fixed length, that feature depending entirely upon circumstances. By much questioning and comparison it would appear that the average nupka was about equal to a quarter of an acre. The squashes were planted in little separate patches of which no account was taken in the land measurement and the sunflowers always went around the outside edge of the whole garden.

The gardens were usually added to year by year until they reached the maximum workable size. Each mature woman in the family usually had her own separate garden where she was helped in her work by her daughters. The smallest garden occupied about three or four nupkas and the largest about nine or ten, or two and one-half to three acres — quite a respectable field to till with the rude implements they had at hand.

Maximilian says (p. 276): ''. . . each family prepares 3, 4, or 5 acres.''

In a family with two or three wives it is quite probable that some four or five acres would be under cultivation, which would yield no mean amount of produce in a good year; and when we remember that each family had at least one and frequently two or more women gardeners, it is evident that there must have been quite an acreage of tilled land in the neighborhood of each village. This is borne out by the accounts of early travelers. Often there was not sufficient available land near the village and part of the gardens were some distance away in the Missouri River bottoms. The Otoes, Missouris, and Pawnees, who lived on the Platte and its forks, had less land suitable for tilling by Indian methods, and their little patches were scattered along the small creek bottoms for miles around the villages. Irving (1833) states that some of the Grand Pawnee corn-patches were eight miles from the village, and that the women took great risks in going out to work in these distant fields. In fact, it was one of the commonest events of Pawnee life for women to be killed in their gardens by lurking Sioux or other enemies. Their bodies were sometimes not found for weeks afterward.

Long's Expedition, 1820 (v. ii, p. 216): "The three Pawnee villages, with their pasture grounds and insignificant enclosures, occupy about ten miles in length of the fertile valley of the Wolf River. The surface is wholly naked of timber, rising gradually to the river hills, which are broad and low, and from a mile to a mile and a half distant. The soil of this valley is deep and of inexhaustible fertility. . . They plant corn and pumpkins along the sides of deep ravines, and wherever by any accident the grassy turf has been eradicated."

Brackenridge, p. 116, says of the Arikara: "Around the village there are little plats enclosed by stakes entwined with osiers, in which they cultivate maize, tobacco, and beans; but their principal field is at a distance of a mile from the village to which such of the females, whose duty it is to attend to their culture, go and return morning and evening."

In order to give some idea of the areas planted about the different villages, the following extracts are given:

Catlin, p. 224 (Arikara): ". . . and we trudged back to the little village of earth covered lodges, which were hemmed in, and almost obscured to the eye, by the fields of corn and luxuriant growth of wild [?] sunflowers, and other vegetable productions of the soil. . ."

Lewis and Clark, p. 158 (Arikara island village): "The island itself is three miles long, and covered with fields in which the Indians raise corn, beans, and potatoes."

Catlin, p. 210 (Hidatsas): "The principal village of the Minitarees, which is built upon the bank of Knife River, contains 40 or 50 earth covered wigwams, from 40 to 50 feet in diameter, and being elevated overlooks the other two, which are on lower ground and almost lost amidst their numerous corn fields."

Henry, p. 323 (Mandans, etc.): "Having got through this wood, we came to the several plantations of corn, beans, squashes, and sunflowers."

Maximilian, p. 20 (Mandans): "We proceeded up the Missouri in a direction parallel with the river, leaving Mih Tutta Hangkush (a village) on our right hand, and taking the way to Ruptare (another village), which runs along the edge of the high plateau below which there is a valley extending to the Missouri, covered with the maize plantations of the Mandans.''

We have already quoted from Matthews, Morgan, Hayden, and Dunbar, and each speaks of the fields being often enclosed with brush fences. Other writers state that the fields were not fenced, so it would appear that fencing depended on the location of the fields and their accessibility to the roving bands of Indian ponies which had free-run of the country for miles around each village. Scattered Corn says that they usually surrounded the gardens with a fence of brush, as did also the men their little tobacco gardens which were always separate from the women's gardens.

Boller also mentions fences: ''The snow had entirely disappeared, except in a few sheltered places amongst the hills, and the Indian women were gathering willows to repair the fence around their corn fields, preparatory to breaking the ground for the coming crop'' (p. 301). And again: ''After the corn is planted and begins to come up, slender fences of willow are necessary to prevent the horses from destroying the slender blades. These willows have to be carried on the backs of the women a long distance, a few at a time, until a sufficient quantity for the purpose is collected'' (p. 118).

Hayden (p. 352): ''These Indians (Arikaras) cultivate small patches of land on the Missouri bottom, each family tilling from one-half to one and one-half acres, which are separated from each other by rude brush and pole fences.'' Morgan, in *Beach's Indian Miscellany*, p. 565, mentions that the Hidatsas surround their garden with similar fences ''of hedge and stakes.''

In the early 70's there were about 400 Mandans, 400 Hidatsas, and 1,500 Arikaras at Fort Berthold. In 1872 the agent reports that these Indians cultivate about 1,000 acres, more than ever before.[14] In 1878 the agent reports 1,292 people (not counting 108 Hidatsas at Fort Buford), families working 295, acreage 800, corn 15,000 bushels, wheat none, vegetables 3,913 bushels, oats and barley 960 bushels; support from agriculture 15 per cent, from hunting, 10 per cent, from the government 75 per cent. At this time game was rapidly disappearing.[15]

Dunbar states that before the government began to plow the lands and encourage the Indians to cultivate the soil on a larger scale, the Pawnees tilled from 1 to 3 acres per family, the average being nearer to 1 than to 3 acres. Their little patches were situated on the creek bottoms, sometimes five or more miles from the villages,

[14] In the early 70's the Indian Office issued instructions to its agents to "favor" the families of Indian men who could be induced to work. Some agents issued more supplies to the men who worked, while others went farther and cut off all supplies from the men who refused to engage in agriculture. This policy resulted in a large increase in the acreage planted by some of the tribes.

[15] Also at this time the Fort Berthold Indians (Mandans, Arikaras, and Hidatsas) had reached about their lowest mark in physical and moral degeneration, due largely to being crowded together in the village with no incentive to hunt and with a large number of the most vicious white men among them. A few years later the government compelled the breaking up of the old village and established each family on its own piece of land.

and were often surrounded by rude fences of interwoven bushes and tree branches.

In 1857, before the government began to plow their lands for them, the Pawnees had about 3,200 people and planted 500 acres of corn (agent's report for 1857). The report for 1867 states that this tribe has about 1,000 acres in cultivation, mostly in corn, with squashes and beans in small patches here and there. In 1878 the Pawnees had 1,292 people on their reservation in Nebraska; besides working 1,000 acres of agency land for the government, the Indians cultivated in small patches 960 acres of corn and secured 8,000 bushels (a poor crop). They also had about 50 acres in squashes, beans, or vegetables. Support from agriculture 30%, from hunting 10%, from the government 60%.

The Ponkas, never much inclined to till the soil, appear to have given up agriculture altogether for some years during the early 60's. In September, 1864, after a successful hunt, they went to the Omaha village to trade buffalo meat for corn (report of the agent for the Omaha tribe, 1864). In 1867 the Ponkas were moved to a new agency on the Niobrara and took up agriculture again. The report for this year states that there were 980 people, that 500 to 600 acres were in cultivation, and that an abundant crop of 13,000 bushels of corn and a large quantity of vegetables were raised. The report for 1874 gives the number of Ponkas as 730; acres cultivated by Indians 300; crops very poor. In 1877 this little tribe was removed to Indian Territory, where they at once succumbed to the chills and fever which always attacked northern Indians in that climate.

After a bitter struggle, during which, to their lasting credit, the whites of Nebraska took the part of these Nebraska Indians, the government was compelled to permit the Ponkas to return to their old home on the Niobrara.[16] Meantime, in the Indian Territory, the tribe had again given up agriculture for some years.

The government began to encourage agriculture among the Omahas early in the 60's. In 1867, 400 acres were broken and planted, besides which the Indians had their own small patches, as to the extent of which the agent gives no figures.[17] The Omahas, always good gardeners, took up the new work very willingly and made surprising progress. In 1878 the statistics show a population of 1,100; acreage 2,200; wheat 21,000 bushels; corn 32,000 bushels; vegetables 7,000 bushels; other crops 1,000 bushels; support from agriculture 95%, from hunting 5%, from the government, none. In this same year the Ponkas, who had been removed to the Indian Territory early in 1877, received 100% support from the government.

Of the Otoes, their agent reports in 1873 that prior to that year their agriculture was still on a very primitive basis, and that they depended mainly on the hunt for their support. The fact seems to be that this tribe had been greatly demoralized through the government's

---

[16] Part of the Ponkas remained of their own choice in the south, and are still living in Oklahoma.

[17] The Sac and Fox agent in his report for 1864 states that the ground cultivated by this tribe is $\frac{9}{10}$ in corn and $\frac{1}{10}$ in beans, squashes, pumpkins, and other vegetables. As far as may be judged, the ratio of corn to vegetables was about the same among the other tribes of the Upper Missouri.

practice of paying annuities in money, most of which the Indians spent for whiskey. In 1857 the Otoes grew about 1,000 bushels of corn in small patches along the bottomland, in the bends of the creeks. They also grew a large quantity of beans and of pumpkins in these patches. The report for 1864 gives us a good idea of the agriculture of the Otoes, as it was before the government began to hire white men to plow the Indian lands. According to this report there were about 500 people on this reservation (Otoes and Missouris); and they had about 140 acres of corn in small patches on the prairie and 100 acres more of corn, beans, squashes, pumpkins, etc., in small patches along the creek bottoms. Both the corn and vegetable patches were worked with hoes. The agent reports in 1874 that until 1873 the Otoes and Missouris had no lands fenced; their small patches were on the bottomlands in the bends of the creeks, and some more on the prairie. This year they had in their own small patches: 200 acres of corn, 15 acres of potatoes, and 10 acres of beans. By 1878 these little tribes were gaining 75% of their support from agriculture and 25% from the government.

Of the Iowas their agent reports in 1859: "This tribe have sixty-eight fields and patches (population 431); the greater part has been cultivated this year." There were about 600 acres in the patches, of which about 500 were planted to corn, wheat, potatoes, beans, pumpkins, and vegetables. The men refused to work. In 1864 there were 293 Iowas on the Great Nemaha Reserve, in southeastern Nebraska; they had 34 "farms," 289 acres in all; this land was plowed by the agency

farmer but planted and cultivated by the Indians, who
raised 6,500 bushels of corn, although it was a drought
year.   Forty-one of the 78 Iowa men were serving in the
Union army at this time.   Their agent in 1867 reports
the "usual acreage" (about 400) of which he says 20
acres were in beans, squashes, pumpkins, and melons.
In 1878 there were 213 Iowas, cultivating 750 acres;
crops, corn 32,000 bushels, vegetables 2,000 bushels; support from agriculture 75%, hunting 1%, from government 24%.

The Osage agent reports in 1867 that this tribe de-
pends mainly on the hunt for support; the women plant
small patches of corn and vegetables and hoe them over
before going on the summer buffalo-hunt; they eat most
of the corn "while soft," but some cache part of it.
The report for 1872 gives some interesting figures on
this tribe:

		Pop.	Acres
	Big Hill's Band	936	125
	Clammore's Band	239	30
Grand Osages	Big Chief's Band	698	None
	Black Dog's Band	511	25
	White Hair's Band	362	250
	Beaver's Band	237	250
Little Osages		696	500
Half-breed Band		277	820

Leaving out the half-breeds, we have 3,679 people and
1,180 acres.

Another interesting tribe of the Lower Missouri region, the Kansa or Kaws, like their close kindred, the
Osages, appear to have taken little interest in agriculture as long as buffalo were abundant.   In 1864 their

population is reported as about 700 and acreage 300, mainly in corn. In 1867 they had 300 acres in cultivation and raised 5,800 bushels of corn, besides vegetables. Population 658. In the early 70's this tribe was removed from Kansas to the Indian Territory, where they joined their kinsmen, the Osages. In 1878 they had 400 people, cultivating 960 acres; crops, corn 8,000 bushels, vegetables 830 bushels; support, from agriculture 75%, from hunting 25%, from government, none.

According to our information, the Indians of the Upper Missouri, while their agriculture was still in its primitive state, cultivated from ⅓ to 1 acre for each person in the tribe; the tribes more engaged in hunting, like the Kansa and Osages, cultivating about ⅓ of an acre for each person in the tribe, while the tribes that depended less on the hunt cultivated about 1 acre to each man, woman, and child. Taking a family as six persons, we would have for the backward tribes 2 acres per family, and for the advanced tribes 6 acres per family. F. A. Michaux, the French botanist, states that in 1802 the American settlers along the Ohio river, using horses to plow their lands, cultivated only 8 to 10 acres per family, and the families were very large ones. He attributes the small acreage planted to the indolence of the people and the attraction of the hunt, game being very abundant. Several other early travelers bear out Michaux's statements as to the very small acreage planted by the American backwoodsmen, who depended partly on the hunt as a means of support. We have similar accounts of the French who settled on the Missouri River above St. Louis after the year 1765. If we go farther back, we find that the Celtic and Anglo-Saxon farmers of England, although they used oxen and plows in their work, did not cultivate much more ground than did the Indian women of the Upper Missouri, and sometimes even less. Thus in one manor near Peterborough (twelfth century) there were 49 men, 20 "full villeins" and 29 "half villeins," with 16 plows; and this manor had only 68 acres in all under the plow.[18] The old acre of Celtic Wales was called the Erw and was 3,413 square yards. Four Erws formed a Tyddyn ("man's house" or homestead), each little farm therefore being less than three modern acres.[19]

---

[18] Robinson, *Readings in European History*, v. i, English Manors.

[19] Skene, *Celtic Scotland or Ancient Alban*, v. iii, p. 22.

III — HARVEST

1. *The return from the summer-hunt*

It was the custom of most of the agricultural tribes of
the Missouri River country to abandon their villages and
crops, as soon as the patches had been given one or two
hoeings, and to go on the summer buffalo hunt. We
know that the Hidatsas and Arikaras went on extended
summer hunts in early days; but by the year 1815,
weakened by smallpox and hemmed in by the Sioux,
they were compelled to give up these tribal hunts as a
regular practice. Now and then when the Sioux were
busy elsewhere or a brief truce was made, these tribes
were free to hunt at a distance from their villages; but
such occasions were rare in later times. The Ponkas,
Omahas, Pawnees, Otoes, Missouris, Kansa, and Osages
continued to go on the summer hunt each year until the
buffalo were destroyed in the 70's. The women usually
gave the patches two hoeings before the tribe started
on the hunt, but sometimes when the season was late
the corn was only hoed once.

Dunbar states that the Pawnees usually set out on
the summer hunt in the end of June; but in later years,
after 1860, owing perhaps to the growing hostility of the
Sioux and the scarcity of buffalo, this tribe often re-

MANDAN SOFT RED CORN

mained at their villages until the middle of July. In 1867 they set out on the hunt July 19th.

Fletcher and La Flesche (Omahas): "When the crops were well advanced and the corn, beans, and melons had been cultivated for the second time, the season was at hand for the tribe to start on its annual buffalo hunt. Preparations for this great event occupied several weeks, as everyone — men, women, and children — moved out on what was often a journey of several hundred miles. Only the very old and the sick and a few who stayed to care for and protect these, remained in the otherwise deserted village. All articles not needed were cached and the entrance to these receptacles concealed for fear of marauding enemies. The earth lodges were left empty, and tent covers and poles were taken along, as during the hunt these portable dwellings were used exclusively" (p. 276).

Dougherty, in *Long's Expedition*, 1819, states that in June an important man among the Omahas would call a council at his lodge to decide whether the hunt should be made toward the northwest, along the Niobrara, or southwest along the upper Loup or Platte. This council also discussed the question as to whether there was a sufficiency of provisions on hand to justify the tribe in remaining at the village long enough to weed the corn.[1] After the council had broken up, criers were sent through the village to call out the decision of the chiefs and to announce the date of departure on the hunt (p. 292).

[1] In several of these early accounts of Indian agriculture on the Upper Missouri, the second hoeing is referred to as "weeding the corn." The hilling-up of the corn also was accomplished during the second hoeing.

Occasionally old or sick persons remained at the village, but usually the entire tribe went on the hunt. The deserted village was often plundered by hostile war parties, who broke into the caches and sometimes damaged the growing crops. The Pawnee caches were sometimes plundered by the Omahas and often by the Otoes, returning famished from an unsuccessful hunt. Pilfering from the caches by neighboring tribes was usually passed over lightly, as the Indians believed that hungry people had a right to take food wherever found; but if too many caches were entered or articles other than food taken, the injured tribe sometimes attacked the guilty one and war followed. The Pawnee caches were most often plundered by the Otoes, an ungrateful little tribe that lived on the lower Platte under the protection of the strong Pawnees. Once the Otoes set fire to the Grand Pawnee lodges while that tribe was on a hunt and burned the village to the ground. Such events, however, were rare, and but little damage was usually done to the deserted villages, and still less to the growing crops, as the small patches were often hidden away in the bends of the creeks miles from the village.

The summer hunt commonly lasted through the hot months of July and August, the tribe returning toward its village early in September. As the time for harvest approached, runners were often sent to the home-village, to examine the crops and report on their condition. Often the tribe was far out in the Plains, or enemies were lurking near the hunting-camp and it was not possible to send runners to the village. On such occasions other means were resorted to. Thus Dunbar informs us that the Pawnees could determine the condition of their corn by examining the seed pods of the milk weed, and when these pods had reached a certain state of maturity the tribe left the hunting-grounds and returned home to harvest the crops.

Dougherty states that the Omahas returned toward their village in August, stopping on the way at the Pawnee villages, to trade guns for horses, and reaching home early in September.

2. *The green-corn harvest*

Among the Mandans, Hidatsas, and Arikaras the first garnering of the harvest started early in August, when the young squashes were gathered, sliced, and dried. At about the same time the green corn season arrived and the corn harvest began. This period usually lasted from the 10th to the 15th of August, but was sometimes earlier. Among most of the tribes the beginning of the green corn season was solemnly determined and announced by some of the more prominent of the older women or by the medicine-men. The elderly women were very expert in determining the condition of the corn from its appearance. Buffalo Bird Woman (Hidatsa) states that the women knew which ears were right for plucking by the dry brown tassels, the dry silks, and the dark green husks — they did not have to open the ear and "look at its face" to see that it was in good green-corn condition, as the educated Indian girls of the present day have to do. She also states that green corn was plucked until frost fell, after which the corn lost its fresh taste; but there was a method for treating the frosted ears that restored much of their original flavor.

The elaborate ceremonies which marked the opening of the green corn season among many tribes appear to have been lacking among the Upper Missouri Indians.[2] The time, however, was one of great rejoicing and feasting, a large part of the crop being consumed during these few days. As Scattered Corn says, they ate just as much as they could during the time when the corn was good.

Not only was this a season of feasting and joy, but also the beginning of the harvest labors, and of the storing of the winter food. The two things went on side by side, as there was always corn boiling and roasting and all were free to help themselves from the ears the women were preparing to dry.

Catlin (p. 212): "The green corn is considered a great luxury by all these tribes. . . It is ready for eating as soon as the ear is of full size, and the kernels are expanded to their full growth, but are yet soft and pulpy."

Catlin (p. 137): "The green corn season is one of great festivity among them and one of much importance. The greater part of their crop is eaten during these festivals and the remainder is gathered and dried on the cob before it is ripened." This is an over-statement of the case, as Scattered Corn says that where a family had a total of nine to twelve nupkas, some two to three nupkas were used up for the feasting and the preparation of dried green corn for winter use. The rest of the crop was permitted to mature.

---

[2] Clark, in the *Indian Sign Language*, states that the Hidatsas had a green-corn dance or ceremony, but he is evidently in error. The Sioux of the upper Mississippi had such a ceremony, which is described in Schoolcraft, *Journal of Travel*, 1820, p. 319.

Boller (p. 118): "Fires are blazing in all directions around which gather merry groups to feast on boiled and roasted ears."

The men sometimes helped at this season, but their labors were not very arduous. Bands of young men went from one field to another, helping to pick the ears, but distinguishing themselves more at the feast, which was always provided for them, than in actual work. It was in the nature of a lark for them.

This green corn which was picked while in the milk and still soft is invariably called "sweet corn" by the early travelers, fur-traders, and frontiermen. It really was the common species of soft field-corn, known as flour corn or starch corn. The Upper Missouri Indians rarely picked the true sweet corn (sugar corn) while green, but permitted it to ripen. The Mandans, Hidatsas, and Arikaras used the sugar corn almost exclusively for the making of corn balls. The tribes farther south, in Nebraska, appear to have used the sugar corn both when green and when ripe in later years; how they used it in earlier times is not known, but as they planted the corn in May and did not return to their villages until early in September, the sugar corn must have been ripe and hard by the time they returned from their summer hunt. Ponka sugar corn planted at Omaha, May 11, 1916, was in green corn condition about August 10 and ripe August 25.

The preparation of this so-called "sweet corn" has been described by many writers.

Boller (p. 118): "Each family reserves a number of the choicest ears to make sweet corn for winter use. It is first parboiled, the grains are then carefully picked off the cob, and dried in the sun upon a piece of lodge skin, prepared thus it retains all its juices and flavor and will keep unimpaired almost any length of time. It is then put away in skin bags and carefully hoarded for use on special occasions or in time of scarcity.''

Hayden (p. 352): ''When green, a portion is gathered, partially boiled, after which it is dried, shelled and laid aside. This is called sweet corn, and is preserved any length of time, and when well boiled it differs very little from green corn fresh from the stalk.''

Report of the Indian agent at Fort Berthold, 1878, p. 32 (Mandans, Hidatsas, and Arikaras): ''They roast great quantities of green corn for winter use by making a long, flat pile of brush, covering it with the corn in the husk, and then burning away the brush. When thoroughly cooked, the burnt husks are removed, the corn shelled and dried and put away. They also dry the squashes for winter food.''

The Upper Missouri tribes prepared this ''sweet corn'' for winter use in two ways: by boiling it in kettles, and by roasting it in fires. The latter method was apparently the older one, and was probably in general use among the tribes before large metal kettles were procured from the white traders.

According to Dunbar the Pawnees returned from their summer hunt about September 1, and at once began to roast and dry the corn that was in the milk. From morning to night the women and children were in the patches, gathering fuel, making fires, picking, roasting, and drying corn. This was called the roasting-ear

time. "In one direction squaws are coming in stagger-
ing under immense burdens of wood and leading lines
of ponies equally heavily loaded. In another the store
of wood is already provided, the fires brightly burning,
in them corn roasting, and near by other corn drying,
while children passing busily to and fro are bringing
loads of corn from the patch. The atmosphere is sat-
urated with the pleasant odor of the roasting and drying
corn."

Continuing his account, Dunbar states that the green
corn, still in the husk, was thrown upon beds of glowing
coals and left to roast. The husks were next stripped
from the roasted ears and the kernels removed from the
cobs. Clam shells were used for this purpose. The
corn after being removed from the cob was spread out
on skins or blankets to dry in the sun, after which it
was packed in rawhide bags and stored away for winter
use. "When roasted in this way," says Mr. Dunbar,
"the corn seems to retain a fineness of flavor which is
quite lost when cooked after our method." [3]

Dougherty states that when the Omahas, returning
from their summer hunt, arrived within two or three
days' journey of their village they dispatched runners
on ahead to ascertain the safety of the village and the
condition of the corn. "On the return of the nation,
which is generally early in September, a different kind
of employment awaits the ever-industrious squaws. The
property buried in the earth [*i.e., cached*] is to be taken
up and arranged in the lodges, which are cleaned out and
put in order. The weeds, which during their absence,

[3] Dunbar, in *Magazine of American History*, v. iv, p. 277.

have grown up in every direction through the village, are cut down and removed.

"A sufficient quantity of sweet corn is next to be prepared for present and future use. Whilst the maize is yet in the milk or soft state, and the grains have nearly attained to their full size, it is collected and boiled on the cob; but the poor who have no kettles, place the ear, sufficiently guarded by its husk, in the hot embers until properly cooked; the maize is then dried, shelled from the cob, again exposed to the sun, and afterward packed away for keeping, in neat leathern sacks. The grain prepared in this manner has a shrivelled appearance, and a sweet taste, whence its name. It may be boiled at any season of the year with nearly as much facility as the recent grain [i.e., fresh green corn], and has much the same taste."[4]

Buffalo Bird Woman states that the Hidatsa green corn season began early in the harvest moon, about the second week of August. The women went to the fields at sunset and each plucked about five baskets of ears and left them in the field over night, to keep fresh. In the morning the corn was carried to the lodge and was there husked, the ears being laid out in rows on the pile of fresh husks. They were next dropped into a kettle of boiling water, a few at a time, and when about half cooked were taken out in a large spoon made of Rocky Mountain sheep horn. These cooked ears were laid in rows on the floor of the drying stage and were left over night. The following morning a lodge cover was spread out, and sitting on this the woman *shelled*

---

[4] Dougherty, in *Long's Expedition*, Thwaites edition, v. i, p. 302.

Above: CROSS SECTION OF AN EAR OF MODERN 24-ROW
WHITE DENT AND OF AN EAR OF 10-ROW MANDAN
WHITE FLOUR CORN

Below: 1. PONKA RED FLOUR. 2. PONKA GRAY FLOUR.
3. PONKA RED SPECKLED FLOUR. 4. PONKA SWEET
CORN

the corn from the cobs, using a mussel-shell for this purpose. Skins were now spread on the floor of the drying stage, and the shelled green corn was spread out on these skins to dry. The corn was permitted to dry for about four days and was then sacked and cached. She also states that the men were away on the "harvest hunt" during the early green corn season, but returned with their meat in time to take part in the later ripe-corn harvest.

At the time when the green corn season began in August groups of young girls sat all day long on the watching-platforms in the center of the corn patches, guarding the corn. At this season the crows and blackbirds were very troublesome, and small boys lurked in the tall weeds, watching for an opportunity to steal the green ears, which they took into the woods and roasted over little fires.[5]

The agent for the Red Cloud Sioux states in his report for 1873 that the corn of the Arikaras is in roasting-ear condition six weeks from planting.

It would be interesting to know just how the Indians hit upon this method of cooking and drying a part of their corn while still green. Featherstonehaugh gives us a clue when he states that the tribes of the upper Mississippi were compelled to cure a large part of their corn in this manner because of the depredations of blackbirds, which attacked the crop in huge flocks at the time when the corn was in the milk.[6]

---

[5] Buffalo Bird Woman's Story, part iii, in *The Farmer*, St. Paul, Minn., Dec. 16, 1916. Same, part iv, Dec. 23.

[6] Featherstonehaugh, *A Canoe Voyage up the Minnay Sotor*, London, 1847. Schoolcraft, *Journal of Travel*, 1820, states that

3. *The ripe-corn harvest*

The green corn season seldom lasted over ten days, after which there was a lapse of time, perhaps two to four weeks, before the actual harvest, of the ripe crop, commenced. The harvest time was perhaps the most important season of the year, and most of the village activities were for the time being subordinated to the interest of getting in the crop.

The Upper Missouri corn was usually ripe and dry in September, though the actual harvest often did not take place until early in October. Maximilian gives the ninth month, September, as the moon of ripe maize. He says, however (p. 276): "The harvest takes place in October when men, women, and children all lend a helping hand."

Dougherty states that the Omahas return to the village early in September to harvest their green corn and ripe corn, and that after the final harvest they leave on their fall hunt late in October. Dunbar gives similar information concerning the Pawnees — that the ripe corn was not gathered until October and that the tribe set out on its fall hunt at the close of the month. The writers who give August and September as the harvest months are evidently referring to the green corn harvest.

When the corn was ripe all of the workers repaired

in the last days of July the corn at Fort Snelling at the mouth of Minnesota River was most of it too hard for eating and part of it hard enough for seed. He also states that the Sioux of Little Crow's village farther down the Mississippi were boiling green corn and holding their green corn ceremonies on August 2, 1820.

BASKETS OF THE MANDANS, HIDATSAS, AND ARIKARAS
The large ones are corn-carrying baskets

to the fields with large carrying-baskets.[7] They went down the rows, snapping the ears and throwing them into small piles at regular intervals. When the time came for returning to the village the ears were placed in the baskets and were carried and piled near the drying platforms in front of the lodges. The next day the women gathered at the piles of snapped corn and the husking began. The husks were stripped back and the outside one removed, leaving the inner and more pliable ones on the ear. The ears were then laid out in rows. The nubbins and poor ears were entirely husked and spread out to dry on one of the floors of the drying stage. Whenever an exceptionably good ear, ripe, and hard, long, straight-rowed, and of good color, came to hand it was stuck into a bag reserved for the seed corn, which was later plaited into separate braids.

After the ears were all prepared in this manner they were braided into strands. There was a standard size for these braids, the length being from knee down around the foot and up to the knee again. One of these braids was the standard measure for corn on the ear; shelled corn was measured by the basket, or by the bullboat load. Among the Hidatsas the standard size of

[7] Skidi Pawnees: ''They also made baskets, which served for various purposes; those used for the transportation of the crops from the fields being especially well made and of unusual construction. Such baskets are in use to-day among the Arikara and Mandan.'' Dorsey, *Traditions of the Skidi Pawnee*, p. xviii.

Arikaras: ''They have preserved in their basketry a weave that has been identified with one practised by former tribes in Louisiana — a probable survival of the method learned when with their kindred in the far Southwest.'' *Bulletin 30*, Bureau of American Ethnology, v. i, p. 85.

the braids of corn was 55 or 54 ears, and in trading with other tribes one of these standard braids was reckoned as the equivalent of one tanned buffalo robe.[8]

Buffalo Bird Woman gives a very interesting account of the Hidatsa ripe-corn harvest. She says that the women went into the fields and snapped the ripe ears, throwing them into piles. This task occupied one day, in the larger fields sometimes two days. On the following morning the women built fires near the piles of corn and set over the fires kettles containing meat and sweet corn. The young men who had been invited to assist at the husking now came into the fields, and a few of the older men usually came also. The girls and young men wore their best clothes and were all painted. The handsome girls always had a large group of these young men gathered about their corn piles.

In husking the corn, any unripe ears that were found were laid aside and became the property of the young man who found them, and he either ate them himself or fed them to his pony. If placed in the caches these unripe ears rotted and spoiled much corn. As they were husked, the ripe ears were thrown into piles, the large ears in one pile, the small ones in another. The large ears were braided in the field by the women, 54 or 55 ears to a braid. After completing a braid a woman often took hold of it by both ends, placed her foot on its middle and gave a sharp tug. This tightened the braid and gave it a neater appearance. The braids of corn were loaded on ponies, about ten strings making a

---

[8] This information on braided corn was gained during recent visits to the Mandans and Hidatsas. — George F. Will.

load, and were taken to the village and hung up on the
rails of the drying stage. The small ears were later
carried to the village in baskets and were spread out on
the floor of the drying stage. The latter task occupied
the women one or two days.

This ripe-corn harvest last about ten days. The young
men worked well, husking out the corn of one field and
then passing on to the next.

After the corn was all braided it was hung about the
frame of the drying scaffold. These scaffolds were an
important feature in the economy of the village, being
used for drying meat as well as corn and vegetables.
In summer the women often gathered under the shade of
the platform's arbor-like roof with their work; at other
seasons the firewood was piled here, and in winter the
arbor was often turned into a stable for the horses.
Dorsey describes the Pawnee arbor or drying platform
as follows: "A few paces to the east of the lodge was
to be found a structure, open at all sides, and with a
flat roof of cottonwood boughs, which served both as a
shelter during the summer and as a platform upon
which could be dried corn, pumpkins, etc."[9] Besides
the roof of branches, the arbor had a floor raised two
feet or more above the ground, upon which the people
sat and the corn was spread to dry. The Mandan scaffold had the floor raised five or six feet from the ground,
and this was reached by means of a ladder cut from a
log. The Arikara dry-stage had a number of long poles
placed lengthwise of the scaffold with willows laid across
them, but the Mandan and Hidatsa stages had only two

[9] Dorsey, *Traditions of the Skidi Pawnee*, p. xvi.

long beams with planks laid crosswise. These tribes build two-section and three-section stages. Part of one of these stages was partitioned off with tipi skins during threshing time, to form a booth in which the corn was pounded out. The Hidatsas, however, never threshed on the floor, and in winnowing they did not pour the grain from the stage floor or roof, but raised it high in baskets and slowly poured it out.[10]

Many of the Indians of today erect these old-time stages or arbors in front of their frame houses.

Boller (p. 118): "When the harvest is gathered in the ears of corn are plaited into a tress (like a rope of onions) and hung upon scaffolds to dry. The variegated hues of the often tastefully arranged traces hanging from the scaffolds gives the village a gay and holiday appearance."

Henry (p. 340): Mandans: "Fronting the porch stands a stage about 8 feet high, 20 feet long, and 10 feet broad, for the purpose of hanging up corn to dry in the fall and to dry meat. These stages have a tolerably good flooring, which in the fall is covered with beans to dry; and posts are erected upon them, on the tops of which are laid poles and rafters, to which corn and sliced squashes [11] are suspended in tresses to dry. When the harvest is over this certainly must have a very

[10] Hidatsa informants state that the corn was always husked in the field and never taken to the lodges to be husked. The men's societies went out in bodies to help with the husking; they were invited by the field owner who paid them with a feast.

[11] Large quantities of pumpkins and squashes were dried for winter use:

"The pumpkins are cut in slips, which are dried in the sun,

Above: Arikara woman threshing corn on the
roof of her house

Below: Scattered Corn, her husband, Holding
Eagle, and their family

pretty effect and give quite an appearance of agriculture.''

After the corn on the scaffolds was dry came the threshing. The braided corn was usually stored as it was, in strings, while the poor ears which were dried on the platform floor were threshed. This was done on the floor itself or upon old tipi skins spread on the ground. The corn was beaten with sticks until all of the grain was loosened from the cobs; the cobs were then picked out and the grain winnowed by letting it fall slowly onto a skin from an elevated place, either from the top of the lodge or the floor of the scaffold.

4. Storing the crop

When the drying and threshing were over the storing of the crop commenced. After a certain amount had been stored in the lodge in dressed skin bags and rawhide parfleches, the remainder of the harvest: corn, beans, squashes, and sunflower seeds, was stored in caches.

The caches of the Mandans are described by Catlin

---

and afterward woven into mats for the convenience of carrying.''
Account of the Pawnee Loups, in *Long's Expedition*, v. ii, p. 217.

''The pumpkins they cut into thin slices and dry in the sun, which reduces it to a small size [bulk] and not more than a tenth of its original weight.'' Account of the Republican Pawnees, in *Pike's Expeditions*, v. ii, p. 533.

Dougherty tells us that the Omahas cut their pumpkins and squashes into slices, dried them and wove them into ''networks or loose mats.'' Some of the tribes wove these dried vegetables into strings or braids.

The Mandans sliced the small green squashes which were strung like rings on a string of braided grass or a leather thong.

as jug-shaped. They were 6 to 8 feet deep, held 20 to 30 bushels and had narrow mouths just wide enough for a person to go through. One for the storage of provisions for immediate use was inside the lodge, back of the fireplace, while others were dug outside, near the lodge.

Fletcher and La Flesche thus describe the Omaha caches: ''Near each dwelling, generally to the left of the entrance, the cache was built. This consisted of a hole in the ground about 8 feet deep, rounded at the bottom and sides, provided with a neck just large enough to admit the body of a person. The whole was lined with split posts, to which was tied an inner lining of bunches of dried grass. The opening was protected by grass, over which sod was placed. In these caches the winter supply of food was stored; the shelled corn was put into skin bags, long strings of corn on the cob were made by braiding the outer husks, while the jerked meat was packed in parfleche cases. Pelts, regalia, and extra clothing were generally kept in the cache; but these were laid in ornamented parfleche cases, never used but for this purpose.

''When the people left the village for the summer buffalo hunt, all cumbersome household articles — as the mortars and pestles, extra hides, etc. — were placed in the caches and the openings carefully concealed. The cases containing gala clothing and regalia were taken along, as these garments were needed at the great tribal ceremonies which took place during that period'' (p. 98).

Dorsey says of the Pawnee caches: ''Just inside the

lodge and to the north of the entrance was built, in winter, the sweat lodge, while at the corresponding position on the south side was an excavation used as a storage cellar for provisions, such as corn and meat, intended for service in the near future. The surplus stock of provisions was cached in excavations generally outside and to the north of the lodge." [12]

John T. Irving, who visited the Pawnees in the early 30's, states that their caches held 100 bushels, but this is evidently an exaggeration.

Pawnees: "On the approach of winter they conceal their stores of corn, dried pumpkins, beans, etc., and with their whole retinue of dogs and horses desert their villages. This they are compelled to do from the want of wood, not only for fuel, but for the support of their numerous horses." [13]

In Miss Mary A. Owen's volume of negro folk-tales [14] there is a very interesting description of the method of preparing and caching "sweet corn" employed by the negro women of western Missouri a generation ago. These negroes of the lower Missouri Valley had a good deal of French and Indian blood in their veins, and Big Angy is said to have been the daughter of a negro moth-

[12] *Tradition of the Skidi Pawnee*, p. xvi.

[13] *Long's Expedition*, v. ii, p. 215.

The Indians fed their horses in winter on the small twigs and bark of the cottonwood tree. The large open groves of these trees, such as the Big Timbers of Republican River, and the Bunch-of-Timbers on the Smoky Hill Fork, were favorite wintering-grounds of the Indians.

[14] *Voodoo Tales, as told among the negroes of the Southwest*, by Mary A. Owen, New York, 1893, p. 40.

er and an Iowa Indian father.   Several old negro women
are talking about food, and Big Angy describes her
method of making and caching ''sweet corn'':

''W'en de roas'in'-yeahs (roasting-ears) is in de milk,
me git um, bile um, dig de grains offen de cob wid lil
stick, spread um on de big rush mats me mek' twell dey
dry lak sand, den me dig hole in de ground — *deep*,
put in de mats all round, den tek de cawn, putt um in
de big bag mek outen de eenside bahk o' de linn-tree,
fling dat bag in de pit, putt on de top mo' mat, shubble
on de dirt, smack um down flat.   Dat *cachein'*!''

''Uh-huh! uh-huh! dat de rale Injun way.''

The bag, made of the inner bark of the ''linn-tree,''
in which the corn was packed, was called a ''splint-bag.''

Verendrye (Mandans): ''Their fort is full of caves
well suited to concealment.'' ''They made us understand that they came inside in the summer to work their
fields, and that there was a large reserve of grain in
their cellars.''

Lewis and Clark (p. 200): ''We purchased from the
Mandans a quantity of corn of mixed color which they
dug up in ears from holes made near the front of their
lodges, in which it is buried during the winter.''

Henry (p. 360): ''The village was soon in an uproar,
the women meanwhile uncovering their stores of corn,
beans, etc. It is customary in the fall, after the harvest
when the grain is well dried in the sun, to take it off the
cob, and deposit it in deep pits about the village. These
holes are about 8 feet deep; the mouth is just wide
enough for a person to descend, but the inside is hollowed out any size, and then the bottom and sides are
well lined with dry straw. Such caches contain from

20 to 30 bushel of corn and beans, which are thrown in loose and covered with straw and earth. The ground is of such a dry sandy nature, that grain stored in this way will keep for several years without injury. So numerous about the village are these pits that it is really dangerous for a stranger to stir out after dark.'' Henry was struck with the great store of corn in these villages when he saw the women preparing to trade with the visiting Cheyennes from the plains: ''On our way we observed the women all busy, taking their hidden treasures and making preparations for the approaching fair. I was surprised to see what quantities they had on hand: I am very confident they had enough to serve them at least twelve months without a supply of flesh or anything else'' (p. 360).

Catlin (v. i, p. 137): ''They are packed away in caches — holes dug in the ground some six or seven feet deep, the insides of which are somewhat in the form of a jug, and tightly closed at the top. The corn and even dried meat and pemican are placed in these caches, being packed tight around the sides with prairie grass, and effectually preserved through the severest winters.''

Hayden (p. 352): ''The crops being gathered in, are stored away in the cellars, before alluded to, or buried on the field in different places, in what are called by the Canadian traders caches, so constructed as to be impervious to rain, and so well covered that no one could discover them without a knowledge of their locality.''

Boller (p. 118): ''The trace [*i.e.*, braided corn] is cached — a hole is dug in the ground, usually near the lodge, some 6 or 8 feet in depth, small at the top but widening as it deepens, much resembling a jug in shape, hay is next strewn over the bottom and sides, and when the corn is thoroughly dried it is taken down from the scaffold and packed away. The cache is filled up with hay, dirt is then thrown on and firmly trodden down, and every sign carefully obliterated. Each family has a generous supply of these caches and as they leave their summer village early in the fall for winter quarters, the corn remains undiscovered and undisturbed until they return in the spring.''

The caches did not, however, always prove safe storage places; for Bradbury informs us that on July 15, 1811, the heavy rains penetrated the Arikara caches and spoiled all their supplies, so that they expected to be in want until the new crop could be harvested.

In early times the Quapaws of Arkansas stored their corn in cane baskets and in gourds ''as large as half-barrels.'' The southern tribes appear to have preferred to store their corn above ground, and it was in the south that the primitive form of the modern corn-crib was invented, by Indians.

Among the Mandans, Hidatsas, and Arikaras, the braided corn was placed in the caches first, after the cache had been lined as Henry describes it.   The threshed or shelled corn was then poured in on top of the braided and the cache filled up. Each family usually had several caches filled with corn, besides those containing beans, dried squashes, sunflower seed, and often dried meat and fat. The position of the caches can still be noted in the old village sites and the number of them cannot fail to impress the beholder with the idea

MANDAN CORN

1. Soft red or purple	7. Soft white
2. Hard white	8. Black
3. Blue	9. Society corn
4. Soft yellow	10. Clay red
5. Blue and white speckled	11. Spotted, or mixed
6. Hard yellow	12. Red sweet corn

that extensive crops must have been necessary to fill them
all. The cache was early adopted by the traders and
trappers for purposes of storage and concealment, and no
account of early travel in the trans-Missouri region fails
to mention some use of caches.

5. Yields

The crops garnered by the Upper Missouri tribes were
at times but a poor reward for the labor expended; but
it is certain that the hardy varieties grown rarely, if
ever, entirely failed; for in the worst seasons at least
seed for the next year's planting was saved. The Ponkas appear to have lost their seed more than once, but
this may have been due to their abandoning their village or eating the little corn left by the drought and
grasshoppers instead of saving it for seed. The agent
for the Red Lake Chippewas of western Minnesota states
in his report for 1878 that some of the land on the shore
of Red Lake has been cultivated for thirty to forty
years and that corn has never failed to make a crop.[15]
The two great enemies to agriculture on the Upper
Missouri in early years were drought and grasshoppers.
Matthews (p. 12) says on this subject: "But they are
not always thus fortunate for the soil of their country,
even that of the Missouri bottoms, is not very rich, the
summer season is short, with early and late frosts, the
climate is dry, long drouths often prevail, to guard
against which they have no system of irrigation; and
lastly the grasshoppers, the plague of the Missouri Val-

---

[15] The corn raised here is of the low flint type, very similar
to, and probably derived from, the Mandan corn.

ley farmer, have often devoured the crops that had escaped all other enemies, and left the Indian with little more than seed enough for the coming spring."

Maximilian (p. 238): "In the heat of summer the creeks become dry, and the crops of maize of the Indians, often fail in consequence of the drought."

Desmet (p. 830): "After preparing this land in this manner they had sowed it. Unluckily this year again the spring had been without rain or even dew. Corn and other vegetables were not growing and their hopes of a good crop were fast vanishing again. The Indians were feeling very bad about it."

The agents for the Sioux of Crow Creek and Cheyenne River report that the crops in 1878 were unusually good, the corn yielding 20 bushels to the acre, and from other reports it would seem that the varieties of corn grown by all of the Upper Missouri tribes yielded about the same — about 20 bushels per acre in a fair year.[16]

Sac and Fox, report for 1874: A good year; corn 20 bushels per acre.

Yanktons, 1867, a very good year, corn estimated 30 bushels to the acre, but the agent is evidently referring to the crop put in by white employes, who perhaps did not plant the local Indian varieties of corn.

Santee Sioux, on Bazile Creek, Knox county, Nebraska, 1878: a very good crop, 350 acres of corn, yield 9,000 bushels — $25\frac{5}{7}$ bushels per acre.

The year 1864 was a drought and grasshopper year all along the Upper Missouri, the agents reporting vari-

---

[16] These Mandan varieties have yielded under very favorable conditions over forty bushels to the acre.

ously from one-third of a crop to a total failure. The Iowas saved a large part of their crop by making it into "sweet corn." Here perhaps we have another reason for the Indian practice of curing a part of their crop while in the milk. The corn was not only very good when so prepared, but the method might be employed in saving the crop from drought, grasshoppers, and birds when it would certainly be lost if left to ripen.

IV — CORN AS FOOD

1. METHODS OF PREPARING CORN. 2. UTENSILS

The corn raised by the agricultural tribes of the Upper Missouri was an important source of food not only to the raisers themselves but to the nomadic tribes of hunters who surrounded their villages on all sides, and also to the early white explorers and traders who penetrated this region.

To the agricultural tribes themselves the corn was not only an important article of food from day to day; but the large stores of grain which they usually had on hand rendered the people more or less independent of the wandering buffalo herds, and insured them against famine, especially during the long, hard winters.

There can be little doubt that, in the time before horses and firearms were acquired, agriculture must have played a much more important part in the economy of the Upper Missouri tribes than it did later. Hunting on foot with bows and arrows, spears, and clubs, the kills must have been small. The trap or corral was of course also resorted to, but herds large enough to make this form of hunting profitable probably did not often come near enough to the villages to make the transportation of the meat on dogs and human backs possible. It is hardly likely that the kills in a region continuously occupied by a large fixed population could have been very frequent, nor that the supplies of meat obtained could have furnished even half of the sustenance of the people.

In support of this view we have the statement of the Assiniboin chiefs, made to Verendrye in 1738, to the effect that the Mandans were usually short of fat, a fact that has been attested by several later authorities. We also have the speech of the Panimaha chief, 1724, in which he speaks of the hardships of the women and *children* who, because of the want of horses, are compelled to carry heavy burdens on their backs during the tribal hunts.[1]

We have seen that Verendrye considered that these Indians were amply provided with grain, and that Henry believed their stores of corn and vegetables sufficient to last the people a full year without other food. This opinion was not held by all of their early visitors; but perhaps the men who state that these tribes were poorly supplied with corn did not witness a general opening of the caches, as did Verendrye and Henry. Pike, who declares that the Pawnee Republics had only enough corn to thicken their soup, camped outside the village and only paid a brief visit to the Indians.

It is probable that as long as buffalo meat was plenty and the road to the herds open, the corn was left undisturbed as an insurance against a possible scarcity of meat or a siege by the enemy, although it is likely that there was an abundance on hand in good years to supply all of their needs during the winter and spring. Scattered Corn says that among the Mandans the supply of most families was usually exhausted by warm weather,

[1] See the account of the Bourgmont expedition, in *Margry*, v. vi.

except for the seed corn, but that some families, who raised the most, nearly always had corn left over. Judging from the statements in *Long's Expedition*, 1820, and Parker's *Travels*, 1835, the Pawnees still had corn on hand in June and July and offered it to their visitors. We have several opinions, on this question of the quantity of corn raised, in the writings of the early explorers and travelers; but one thing seems clear, and that is that corn was sparingly used when other food was plentiful, although the Indians might have a large supply on hand for purposes of trade. Meat usually had small value in trade, as compared with the value of corn, and the Indians of the earth-lodge villages appear at most times to have preferred to eat the meat and save most of the corn for purposes of trade. Verendrye, however, seems to imply an entirely different condition in his time, as he states on the authority of the Assiniboins that the Mandans had little meat and fat and lived almost entirely on corn and vegetables.

Lewis and Clark, p. 231 (Mandans): "Their principal article of food is buffalo meat, their corn, beans, and other grains being reserved for summer, or as a last resource against what they constantly dread, an attack from the Sioux."

Yet these explorers themselves used several hundred bushels of Mandan corn during that winter. Perhaps the Indians employed the above explanation as a trader's trick to keep up the price.

Catlin says of the Mandans (v. i, p. 137): "Corn and dried meat are generally laid in in the fall, in sufficient quantities to support them through the winter.

These are the principal articles of food during that long and inclement season; in addition to them they often-times have in store large quantities of dried squashes and dried pommes blanches. . ."

Boller was a later visitor and saw these tribes at the time when they were cut off from the buffalo herds. He says (p. 118): "They raise black beans, pumpkins, and squashes, but in spite of these vegetable resources hemmed in as they often are by enemies, and consequent-ly unable to obtain by hunting a full supply of buffalo meat, they sometimes suffer greatly for food." And on p. 156: "The surplus corn had been cached and the scarcity of buffalo around the summer village began to be severely felt. Had it not been for their crops the Indians would have been reduced to extreme hunger."

From these statements we may suppose that conditions were pretty much the same both in early times and in later years: That these tribes usually had a large sur-plus of corn and vegetables when they had access to the buffalo herds and when friendly hunting tribes came to their villages to exchange dried meat for corn and other articles; but that when the tribes were cut off from the usual supply of meat they consumed all of their store of corn and vegetables, and if the last crop had not been abundant, they were probably brought to the verge of famine.

1. Methods of preparing corn

The importance of corn as a food among the tribes of the Upper Missouri has been already referred to, and some account must now be given of the methods of pre-paring and cooking corn.

The dried green corn, or sweet corn, was usually boiled, sometimes by itself, but more often with beans, with squashes or pumpkins and with the roots or fruits of certain wild plants. The hard ripe corn was either parched, made into hominy, or pounded into meal. Matthews gives a good general statement on the uses of corn among the Mandans, Hidatsas and Arikaras:

"Their principal vegetable diet was the corn they raised themselves. Flour, issued by the Agency is now to a great extent taking its place. They eat some of the corn while it is green, but the greater part they allow to ripen. When ripe they prepare it in various ways; they pound it in a wooden mortar with water, and boil the moist meal thus made into a hasty pudding, or cook it in cakes; they frequently parch the corn and then reduce it to powder which is often eaten without preparation. A portion of their corn they boil when nearly ripe, they then dry and shell it and lay it by for winter use. When boiled again it tastes like green corn; this is often boiled with dried beans to make a succotash. Their beans are not usually eaten until ripe; squashes are cut in thin slices and dried; the dried squash is usually cooked by boiling; sunflower seed are dried, slightly scorched in pots or pans over the fire and then powdered; the meal is boiled or made into cakes with grease. The sunflower cakes are often taken on war parties and are said when eaten even sparingly to sustain the consumer against fatigue more than any other food" (p. 25).

Dougherty (1819) gives a similar account of the uses of corn among the Omahas: The dried green corn "may

MANDAN SQUASH

be boiled at any season of the year. . . They also pound it into a kind of small hominy, which when boiled into a thick mush, with a proper proportion of the smaller entrails [2] and jerked meats is held in much estimation. When the maize which remains on the stock is fully ripe, it is gathered, shelled, dried, and also packed away in leathern sacks. They sometimes prepare this hard corn for eating, by the process of leying it, or boiling it in a ley of wood-ashes for the space of an hour or two, which devests it of the hard exterior skin; after which it is well washed and rinsed. It may then be readily boiled to an eatable softness, and affords a palatable food.

"The hard ripe maize is also broken into small pieces between two stones, one or two grains at a time, the larger stone being placed on a skin, that the flying fragments may not be lost. This coarse meal is boiled into a mush called Wa-na-de. It is sometimes parched previously to being pounded, and the mush prepared from this description of meal is distinguished by the term Wa-jung-ga. With each of these two dishes, a portion of the small prepared intestines of the bison, called Ta-she-ba, are boiled, to render the food more sapid." [3]

Green corn was roasted or boiled on the ear and was sometimes buttered with fat or marrow: "When boiled green with rich buffalo marrow spread on it (instead of

---

[2] The small intestines of the buffalo were cleaned and turned or inverted so as to enclose the fat which covered their exterior surface. They were then dried in the sun, woven into braids or mats, and stored away for winter use.

[3] Dougherty, in *Long's Expedition*, Thwaites edition, v. i, p. 303.

butter), it is very sweet and truly delicious." (Boller, p. 135.)

"With ears of green corn, which they either bury in the embers, still enveloped in the husks, or roast before the fire; when sufficiently done they season it with bear's oil, buffalo suet or marrow, and partake of the rich though simple repast with joyful gratitude." (Hunter, p. 273. This evidently refers to the Osages.)

Sweet corn, or green corn roasted or boiled and then dried in the sun, was called *Watongziskithe* by the Omahas. The corn prepared in this manner for winter use is very highly spoken of by all of the early travelers and traders.

"Our supper consisted of very pleasant flavored sweet maize." (Maximilian, p. 40, — Mandans.)

Boller, p. 135 (Mandans): "The feast consisting of Indian sweet corn and tea was set before them."

"We were conducted to the lodge of one of their chiefs, where there was a feast of sweet corn, prepared by boiling and mixing it with buffalo grease. Accustomed as I now was to the [de]privation of bread and salt, I thought it very palatable. Sweet corn is corn gathered before it is ripe, and dried in the sun; it is called by the Americans green corn or corn in the milk." (Bradbury, p. 131, — Arikaras.)

Sweet corn and beans boiled together in the form of a succotash was a favorite dish among the tribes of the Upper Missouri. It was prepared in various ways.

Henry, p. 327: "Our host presented us with dried meat, and then a dish of corn and beans; but as the latter is not cooked with any kind of grease or fat, it has a very insipid taste."

Maximilian, p. 28 (Hidatsa buffalo dance): "Several young men were now employed carrying around dishes of boiled maize and beans which they placed before the guests. These dishes were handed to each person successively, who passed them on after tasting a small quantity."

According to Carver the Wisconsin Indians prepared succotash from fresh green corn and unripe beans: "One dish which answers nearly the same purpose as bread, is in use among the Outagamies, the Saukies, and other eastern nations, where corn grows . . . is reckoned extremely palatable by all Europeans who enter their domains. This is composed of their unripe corn, as before described, and beans in the same state, boiled together with bear's flesh — which renders it beyond comparison delicious. They call this food succotash" (p. 135). The Mandans, Hidatsas, and Arikaras do not appear to have made succotash of fresh green corn and unripe beans; but Dunbar states that the Pawnees boiled their beans and pumpkins when green, as well as when ripe.[4]

Henry, p. 325: "On going into the hut we found buffalo hides spread on the ground before the fire for us to sit upon, and were presented with two large dishes of boiled corn and beans."

Maximilian, p. 220: "A large dish of boiled maize and beans was immediately set before us. It was very tender and well dressed and three of us ate out of the

---

[4] Buffalo Bird Woman states that the young girls who watched the corn patches to prevent the birds damaging the green corn sometimes boiled green corn and green beans together and ate them in the fields.

dish with spoons made out of the horn of buffalo or big-horn."

To the simple dish of boiled corn and beans, other vegetables, meats, and dried fruits were often added:

Bradbury, p. 154 (Hidatsas): "We were treated with a dish consisting of jerked buffalo meat, corn and beans boiled together."

Long's Expedition, v. i, p. 114 (Kansa tribe): "They commonly placed before us a sort of soup, composed of maize of the present season [August 20], of that description which having undergone a certain preparation [roasting and drying], is appropriately named sweet corn, boiled in water, and enriched with a few slices of bison meat, grease, and some beans, and to suit it to our palates, it was generally seasoned with rock salt, which is procured near the Arkansa river."

Lewis and Clark, p. 214: "Kagahami or Little Raven brought his wife and son loaded with corn, and she then entertained us with a favorite Mandan dish, a mixture of pumpkins, beans, corn, choke cherries with the stones, all boiled together in a kettle, and forming a composition by no means unpalatable."

Fletcher and La Flesche, p. 270 (Omahas): "*Um' bagthe*: corn boiled with beans, set over night to cool and harden then served cut in slices. Considered a delicacy."

Parker's *Travels*, 1835, p. 53 (Pawnees): "The daughters of Big Ax served us on the occasion, and bountifully helped us to boiled corn and beans. . . In the evening we were invited to two other feasts. The first consisted of boiled corn and dried pumpkins boiled together, and the other of boiled buffalo meat."

PAWNEE CORN

1. Blue flour
2. White flour
3. Blue speckled
4. Red striped

5. Yellow flint
6. Red flour
7. Yellow flour
8. Sweet corn

LaRaye, p. 157 (Pawnees): "When they boil it [the meat] they continue boiling it until it can be eaten with a spoon, throwing in a handful of corn if they have it, with a small quantity of bear's oil."

Dried sweet corn, therefore, was boiled by itself, with beans, with pumpkins and squashes, with meat and the small dried entrails of the buffalo, and was sometimes seasoned with salt, with bear's oil, or with choke cherries. The choke cherries were pounded up stones and all, and were then made into thin cakes which were dried in the sun for winter use. Salt was obtained by the Omahas from Salt Creek near the present city of Lincoln, Nebraska, and by some of the other tribes from the salt plains on the Cimarron and Salt forks of the Arkansas.[5]

Parched corn was called *Wana'xe* by the Omahas, and skin bags of this food were carried on hunting trips, war expeditions, and other journeys. It was usually parched by thrusting a sharpened stick into the butt of the ear and holding it over the fire.

Henry, p. 369: They "presented the pipe, some meadow turnips, and a few ears of very hard, dry corn which the women had parched."

[5] Near the Pawnee villages in 1820, Long's party met a young Arikara man who lived with the Skidi Pawnees. "He had brought with him from one of the upper branches of the Arkansas, two masses of salt, each weighing about thirty pounds. This salt is pure and perfect, consisting of large crystalline grains, so concreted together as to form a mass about twenty inches in diameter and six in thickness." V. ii, p. 219.

Dunbar states that in early years the Pawnees carried on a trade in this salt with the tribes farther up the Missouri, including the Mandans.

Dunbar (Pawnees): "The corn was sometimes parched before triturating, and by this means the flavor of the food was much improved."

Catlin, p. 169 (Four Bears' exploit): "He traveled the distance of 200 miles entirely alone, with a little parched corn in his pouch."

Parched corn was often pounded into a coarse meal and boiled into a mush. The Omahas called this *Wa-shon'ge*. "A stick, *nonxpe*, was thrust into the cob and the corn roasted before a fire; then it was shelled and the chaff blown off; finally it was pounded in a mortar (*uhe*) with a pestle (*wehe*)." Fletcher and La Flesche, p. 270.

Corn meal was made both of parched and unparched corn and was much employed in the cookery of all of the Upper Missouri tribes.

Henry, p. 327: "The corn is generally bruised or pounded in a wooden mortar, which is fixed firmly into the ground in one corner of the hut; and this is the first work performed by the women in the morning — after having washed themselves in the Missouri."

Verendrye speaks of the Mandans going to meet him "with coarse grain cooked, and flour made into paste."

Bradbury, p. 133 (Arikaras): "The squaw prepared something for us to eat; this consisted of dried buffalo meat mixed with pounded corn, warmed on the fire in an earthen vessel of their own manufacture. Some offered us sweet corn mixed with beans."

Lewis and Clark, p. 189 (Mandans): "We received a visit from Kagahami or Little Raven, whose wife accompanied him, bringing 60 pounds of dried meat, a robe, and a pot of meal."

Henry mentions fresh corn (green corn?) mixed with parched corn meal: "We were invited into several huts successively and presented with . . . corn and beans, together with parched corn and fresh ears pounded up in a mortar; this last dish we found good" (p. 325). And again, p. 332, he says: "We had a plentiful supply of corn and beans, and were soon invited to several huts, where we were treated with a very palatable dish of pounded peas and parched corn."

Fletcher and La Flesche mention two Omaha dishes made of corn meal: "*Wa'ske*: pounded corn mixed with honey and buffalo marrow. *Wani'de*: mush or gruel — pounded corn mixed with water" (p. 270).

Bread made of corn and beans is occasionally mentioned among the Upper Missouri Indians.

Lewis and Clark, p. 161 (Arikaras): "They also brought us some corn, beans, and dried squashes . . . they presented us with a bread made of corn and beans, also corn and beans boiled, and large, rich beans which they take from the mice of the prairie."

Merrill mentions this same food among the Otoes: "a piece of bread made of pounded corn and beans, baked in the ashes." [6]

Dunbar states that the Pawnees made the meal from the mortar into cakes which were baked in the ashes or on hot stones, and the Spanish accounts of Coronado's expedition, 1541, state that the Quiviras and Harahays (Wichitas and Pawnees?) had no bread, except the kind "baked under the ashes."

Bread made of fresh green corn was made by the

[6] Nebraska Historical Society, *Transactions*, v. v, p. 230.

Hidatsas, according to Buffalo Bird Woman.[6a] She says that the corn was shelled from the cob with the thumb nail and was then pounded to a pulp in the mortar. A row of fresh husks was laid down, overlapping like shingles, and on these several more layers were laid, every other layer running crosswise. The corn pulp was then poured on the husks and patted into a cake about two inches thick. The edges of the husks were next drawn over the top of the cake, and the husk covering was then tied down with strips of husk. A cavity in the ashes of the hearth was made, live coals were raked into this; the cake was placed upon the coals and was covered with more ashes and coals. Two hours were required to bake such a cake. No fat or seasoning was used. This was called Naktsi: Thing-baked-in-ashes.

Corn balls were a favorite article of food among the Mandans and Hidatsas. There were several varieties of corn balls. One was made of pounded sugar corn mixed with grease. Scattered Corn states that the sugar corn was used almost wholly for this purpose. It was never picked green. The Hidatsas give the same information. The corn ball made of sugar corn was called by the Mandans *wüpe*. Another kind of corn ball was made of pounded corn, pounded sunflower seed, and boiled beans.

[6a] Buffalo Bird Woman: All of our quotations from this Hidatsa woman should be credited to Dr. Gilbert L. Wilson, who is the author of the series of articles, ''The Story of Buffalo Bird Woman,'' which appeared in *The Farmer*, a weekly farm journal, published at St. Paul, Minn. This very interesting life story of an Indian woman appeared in four parts in the issues of *The Farmer* for December 2, 9, 16, and 23, 1916.

It tasted like peanut butter and was called *opata* by the Mandans.

Old Cheyennes state that corn balls were an article of food the eagle-catchers always took with them into the eagle-pits. The practices connected with the art of eagle-catching were regulated by a set of very old customs, which might almost be termed a ritual, and the eating of corn balls in the eagle-pits was one of these practices.

Henry refers to corn balls (p. 357): "He presented me with a dish of water, which after my taking a draught, he removed, and handed me a dish containing several balls, about the size of a hen's egg, made of pears [june berries], dried meat and parched corn, beaten together in a mortar."

Henry (p. 400): "We also paid some women for preparing provision for our homeward journey; this was principally parched corn pounded into flour, mixed with a small portion of fat, and made up into balls about the size of an egg. These may be eaten as they are, or boiled for a short time; the latter method we found most wholesome."

Larpenteur, p. 247 (Arikaras): "As we thus drifted along with the current, they gave us some of their provisions, which were little balls, made of pounded parched corn, mixed with marrow-fat, and some boiled squashes."

Hominy was made by several of the Upper Missouri tribes from the hard ripe corn. The best description we have of the method employed in making hominy is contained in the account of the Kansa tribe in *Long's Expedition* (v. i, p. 114): "Another very acceptable dish

was called *leyed corn*; this is maize of the preceding
season shelled from the cob, and first boiled for a short
time in *ley* of wood-ashes until the hard skin, which in-
vests the grains, is separated from them; the whole is
then poured into a basket, which is repeatedly dipped
into clean water, until the ley and the skins are removed;
the remainder is then boiled in water until so soft as to
be edible.''

Dunbar mentions the hominy made by the Pawnees,
but does not describe the process.

The Omahas made from the hard ripe corn *Wabi'-
shnude* — ''corn boiled with ashes and hulled.'' Dougherty describes the Omaha process as follows: ''They
sometimes prepare this hard corn for eating, by the process of leying it, or boiling it in a ley of wood-ashes for
the space of an hour or two, which divests it of the hard
exterior skin; after which it is well washed and rinsed.
It may then be readily boiled to an eatable softness.'' [7]

Lewis and Clark mention hominy among the Mandans:
''In the course of the day we received several presents
from the women, consisting of corn, boiled hominy and
garden stuffs'' (p. 181).

Brackenridge applies the name hominy to the dried
sweet corn, which was never used in making real hominy: ''After the meat they offered us hominy made out
of corn dried in the milk, mixed with beans, which was
prepared with buffalo marrow and tasted extremely
well. . . Their most common food is hominy and
dried buffalo meat'' (p. 116). This is evidently the same
food that Dougherty mentions among the Omahas —

---

[7] *Long's Expedition*, v. i, p. 303.

dried sweet corn pounded into "a kind of small hominy, which when boiled into a thick mush, with a proper portion of the smaller entrails and jerked meat, is held in much estimation." Bradbury also mentions "hominy and boiled buffalo meat" among the Arikaras.

Squashes and pumpkins, both green and dried, were highly esteemed by the Missouri Valley Indians. The use of these vegetables has been already referred to, but a few additional quotations may be given here.

Hayden, p. 352: "After corn squashes next claim their attention in agriculture. They grow on large and strong vines and are of various sizes and shapes. They are either boiled and eaten when green or cut up and dried for winter use. In the latter case they become very hard, and are scarcely eatable when cooked, except by the natives, who seem to devour them with a gusto and preference not shown for any other vegetable except corn."

Catlin refers to the pounding up of the dried squashes into meal: "These are dried in great quantities and pounded into a sort of meal, and cooked with dried meat and corn" (v. i, p. 137).

Dunbar states that the Pawnees grew considerable quantities of pumpkins and squashes, gathered them when they did the green corn in September, cut them into long slices and hung them up on scaffolds to dry. They were eaten both fresh-picked and dried. Dried pumpkin boiled with buffalo meat is mentioned among the Pawnees by Parker, July 3, 1835. This was a favorite dish.

Choke cherries were much employed as a seasoning for

food both by the agricultural tribes along the Missouri River and the wandering hunter tribes in the Plains. These cherries were pounded up stones and all, made into thin cakes and dried in the sun for winter use. The Plains Indians boiled pieces of this cherry cake with their meat, giving the broth a pleasant flavor. The Arapahoes mingled buffalo tallow with the pounded cherries before drying them in the sun.

2. Utensils

In the work of preparing and cooking the food, there were a number of receptacles and utensils in use among the Upper Missouri tribes. Included in this category are the wooden bowls and dishes, clay pottery, baskets, spoons, stone mullers, mortars and pestles.

The wooden dishes and bowls were used to hold the seed in the field while planting as well as in the serving of food in the lodge. They were carved out of some fairly hard wood, usually, and do not seem to have been nearly as common as the clay pottery.

Fletcher and La Flesche, p. 338 (Omahas): "The making of wooden articles was also the task of the men. . . . Wooden bowls (*zhongu'xpe*) were made from the burrs of the black walnut. These were burned into shape as described and polished with sand and water; experience and skill were needed to make the bowl symmetrical. Some of these bowls were beautiful in the marking and grain of the wood as well as in form."

Dorsey, *Traditions of the Skidi Pawnee*, p. 18: The ordinary domestic utensils were the wooden bowl or platter, beautifully carved from a knot, and spoons made of the buffalo horn, in many shapes."

We have many good early descriptions of the manufacture of pottery among these tribes.

Henry, p. 328 (Mandans): "Their corn and beans are boiled whole. They use large earthen pots of their own manufacture of a black clay which is plentiful around their villages. They make them of different sizes from one quart to five gallons. . . One or more of the largest kind is constantly boiling prepared corn and beans, and all who come in are welcome to help themselves to as much as they can eat of the contents. The bottoms of these pots are of a convex shape and care is therefore required to keep them from upsetting. For this purpose when they are put to the fire a hole is made in the ashes to keep them erect, and when taken away they are placed upon a sort of coil made of bois blanc fibers. These coils are of different sizes, according to the dimensions of the several pots. Some pots have two ears or handles and are more convenient than those with none."

Hayden, p. 355 (Arikaras): "The Arikaras though stupid in many respects show considerable ingenuity in making tolerably good and well shaped vessels for cooking purposes. They are wrought by hand out of clay, and baked in the fire, though not glazed. They consist of pots, pans, porringers, and mortars for pounding corn. They are of a gray color; stand well the action of fire, and are nearly as strong as ordinary potters' ware."

Fletcher and La Flesche (Omahas): "In old times the Omaha women made pottery of a rather coarse type, ornamented with incised lines. These pottery kettles could be hung or set over the fire. . . Bowls of pot-

tery and wood were used, which bore the general name
uxpé" (pp. 340-341).

The baskets which were used in carrying to and from
the field and as receptacles about the lodge were woven
in pleasing patterns of strips of box-elder bark, that is
the inner bark, and of the inner bark of a species of willows. The frame was of willow rods bent into shape and
tied with strips of hide. The weaving was neither very
fine nor complicated, but gave a strong and durable basket. The art of making them is almost lost among the
Mandans, Hidatsa, and Arikaras, there being but one
Arikara and one Hidatsa woman who still practice it.
Dorsey tells us that the Skidi Pawnee formerly made
carrying-baskets, such as "are in use today among the
Arikara and Mandan." The Omahas understood the
art of weaving after a rude fashion, but there is no mention of baskets being made or used.

Verendrye (Mandans): "They make wicker work
very neatly, flat and in baskets. They make use of
earthen pots, which they use like many other nations for
cooking their food."

Spoons and ladles were made both of wood and horn.
The horn spoons in particular were very graceful in
form, the smaller ones being made of the black horn of
the buffalo, while the larger ones, often holding a pint or
more, were carved from the yellow horn of the bighorn
or Rocky Mountain sheep.

Fletcher and La Flesche (Omahas): "Wooden ladles
were made with the handle so shaped that it could be
hooked on the edge of the bowl so as not to drop into
the contents. . . Spoons were made of wood and of

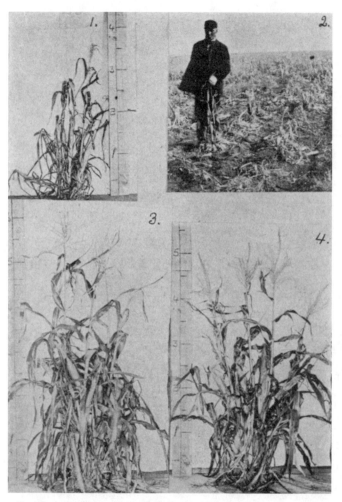

By permission of the Montana Agricultural Experiment Station

PLANTS OF FOUR VARIETIES OF SIOUX CORN

1. Fort Totten Sioux 2. Field of Fort Peck corn
3. Santee Sioux 4. Blue flour from Rosebud Agency

buffalo horn; the latter kind were in general use although tabu to one subdivision of the Tha'tada gens" (p. 338). This subgens held the head of the buffalo tabu and therefore might not touch anything made of buffalo horn.

Each house had its permanent wood mortar, set firmly into the earth floor, with a heavy wooden pestle fitting into it. In this most of the corn grinding or pounding was done. In addition to these wooden mortars each woman had several small stone mortars. Some of these were flat stones; others were of the shape of the stones used in the Scotch game of curling; each stone mortar had a small hollow in the top, and small rounded stones, two to three inches in diameter, were employed as pounders. These little stone mortars were used in pounding up choke cherries and other fruits, dried meat, and hard corn. The hollow in the stone was so small that only three or four grains of corn could be broken at a time. Among the tribes that abandoned their villages and went on extended tribal hunts, the large wooden mortars were taken up and stored in caches and only the small stone mortars were taken on the hunting trips. Among the older people of the Mandans, Hidatsas, and Arikaras, the large wooden mortar and the flat stones are still in use at the present time.

Hayden speaks of baked clay mortars being made by the Arikaras: ''They make of clay and bake in the fire pots, pans, porringers, and mortars for pounding corn. . . For pounding corn and other hard substances, they make also mortars of stone, working the material into shape with great labor and perseverance'' (p. 355).

Fletcher and La Flesche (Omahas): "The making of wooden articles was also the task of the men. The mortar (*u'he*), which was a necessity in every household, was formed from a section of a tree-trunk a foot or so in diameter and about three feet long. One end was chipped to a point so that it could be thrust into the ground to hold the utensil steady when in use; the other end was hollowed out to form the receptacle for the corn, by the following process: Coals were placed on the surface and were kept 'alive' by being fanned as they slowly burned their way into the wood, until a sufficiently large cavity had been burned out, when the mortar was smoothed with sandstone and water, inside and outside. The pestle (*we'he*) was between three and four feet long, large and heavy at one end, and smaller and tapering at the other. When in use the small end was inserted into the mortar, the weight of the large end giving added force to the pounding of the corn" (p. 338).

Dougherty (1819) informs us that when the Omahas started on their annual buffalo hunt they concealed these large wooden mortars and pestles in caches and took only the small stone ones with them.

V — CORN AS AN ARTICLE OF TRADE

1. EARLY INTERTRIBAL TRADE. 2. TRADE WITH THE
WHITES

1. Early intertribal trade

Like the towns of Pecos, Taos, and Picuris on the eastern and northern frontiers of the Pueblo country, the earth-lodge villages of the Upper Missouri were from very early times centers of trade to which the wandering tribes in the Plains came at frequent intervals, to visit and traffic with the inhabitants. In primitive times these hunters of the Plains brought to the Missouri villages meat and robes, pelts, eagle feathers, red pipestone from the Minnesota region, and rock salt from the plains south of the Arkansas. By the beginning of the eighteenth century European goods made their appearance, the tribes of the north bringing down English goods from Hudson Bay to the Missouri, while the Plains tribes brought Spanish goods from New Mexico. It was at about this period also that the Plains tribes began to bring horses to the Missouri villages in considerable numbers, to be exchanged for guns and other articles from Hudson Bay.

In this early intertribal trade the villagers acted as middlemen between the tribes to the north and east of the Missouri and those in the Plains to the south and west. The trade found its way naturally to the villages, because of their fixed position along the great dividing line of the Missouri, and also because of the agriculture practised by the village peoples. The hunters of the Plains, like all people subsisting almost entirely on meat, had a craving for vegetable foods; they had a particular liking for corn; and the villages on the Upper Missouri were the only points between the Mississippi and the pueblos of New Mexico at which corn could be procured. This traffic in corn between the villagers and the men of the Plains is mentioned by the Spaniards as early as 1541, and it continued until the Plains tribes were placed on reservations and rationed by the United States government, fifty or sixty years ago.

The custom of giving presents was universal among the tribes, and as a return was expected this form of exchange was usually as profitable as regular trading. From very early times the agricultural tribes of the Upper Missouri made a practice of of giving large presents of corn to visitors.

Verendrye is the first to mention this custom (1738). He states that he was met near the Mandan villages by a messenger who "presented me with a gift of Indian corn in the ear, and of their tobacco in rolls, which is not as good as ours as they do not know how to cure it like us."

Lewis and Clark make frequent mention in their journals of presents of corn from the Mandans and Hidatsas. "They made us a present of 7 bushels of corn, a pair of leggings, a twist of their tobacco, and seeds of two different species of tobacco" (p. 161). Again, a Mandan sent an invitation "to come to his village, where he wished to present some corn to us" (p. 186). And (p. 181): "They brought a present of corn on their mules."

The agricultural tribes of the Upper Missouri conducted a large trade in corn with the hunter tribes of
the plains. This trade was of great importance to the
agricultural tribes, particularly in early times before
white traders had established themselves on the Upper
Missouri. It was through this trade that the tribes procured their first guns and other European weapons, without which they would have been fairly at the mercy of
their powerful neighbors in the east and northeast, who
were in direct trade relations with the French and English. The corn trade also brought the earth-lodge villagers their first horses, animals which proved of incal-
culable value; and the fact that they always had a surplus stock of corn and vegetables on hand for purposes
of barter frequently won for them, for short periods at
least, the friendship of these powerful neighbors, and
thus gave their warriors seasons of much-needed rest
from the continuous strain of standing on guard against
hostile raiders year after year.

To the hunting tribes the corn trade was also of high
value. Writing in 1826, Thomas Forsyth, one of our
best early authorities on the Indians of the Upper Mississippi and Upper Missouri regions, states that Indians,
even when they had an abundance of meat, always felt at
a loss without vegetable food, and that they had a particular craving for corn.[1] In early times journeys of
surprising length were made by the hunter tribes to
procure corn at the Upper Missouri villages. From the
southwest they came from the headwaters of the Platte,

[1] Thomas Forsyth's account of the Sauk and Fox tribes, re-
printed in Blair's *Indian Tribes of the Upper Mississippi*, v. ii,
p. 228.

and from the northeast from at least as far as the Assiniboin River and the Lake of the Woods. To these
non-agricultural tribes the corn meant in times of plenty
a welcome and needed change of diet; but in hard winters, and when the buffalo or other game eluded their
search, it meant many times the salvation of whole
camps, destitute of meat and perishing of starvation.
Without this stored-up food within reach the toll of
famine would have been immeasurably greater in many
a hard winter. On these occasions horses, arms, and the
stores of furs, which they were better able to gather during their wanderings than were the sedentary village Indians, all went to the corn raisers in exchange for the
saving grain.

We have some long accounts and frequent brief mentions of this trade in corn which demonstrates conclusive-
ly its great importance and the high place which it occupied in the economy of the entire northern plains area.

Going back to the earliest period of which we have
any written record, we find that when the Spaniards
penetrated to New Mexico in 1540 they found that the
Pueblo Indians had a definite knowledge of the Wichitas
and, evidently, of the Pawnees, living in the southern
edge of the Upper Missouri region. The New Mexican
Indians had slaves from these tribes; and as we have no
reason to suppose that they carried their war expeditions
as far as the Pawnee country, we must conclude that
these Wichita and Pawnee slaves found their way to
the Southwest through the medium of trade with the
wandering hunters in the Plains, and that a system of
intertribal trade extending from the Missouri to New

Mexico was already in existence early in the sixteenth century, although at that period the tribes in the Plains were all on foot.[2] It was through this intertribal trade in corn and other articles that the Spaniards in New Mexico during the seventeenth, eighteenth, and early nineteenth centuries secured frequent information as to events on the Missouri River. When the French explorers pushed out into the plains of Oklahoma and Kansas, early in the eighteenth century, they found a knowledge of the wonderful earth-lodge villages of the Dakota region extending as far south as the tribes dwelling on the Canadian River in Oklahoma; and they mention Arikara slaves among the Ietans on the northern frontier of New Mexico. In a similar way, either toward the close of the seventeenth century or early in the following one, the Assiniboins and Crees from the north carried a knowledge of the Mandan villages along the Great Lakes to the outposts of the French and northward to the English on Hudson Bay. The tales they brought, either very meager or misunderstood by the French and English, led to the belief that the Mandans were a white people, perhaps Spaniards, living on or near the shores of the western sea.

What appears to be the earliest mention of this Mandan trade in corn with the northern tribes, is contained in a letter from Father Aulneau, Jesuit missionary among

[2] The chroniclers of the Coronado expedition, 1540-41, state that these wandering foot Indians of the Plains came in near the Pueblo towns each fall and ''wintered under the wings'' of the villages. They were not permitted to enter the towns, as they could never be trusted, but the Pueblos were always glad to trade with them, giving corn for dried meat and buffalo skins.

the Crees, dated April 30, 1736. In this letter Aulneau
speaks of Lake Winnipeg and states that he intends to
spend the coming summer among the Assiniboins who
occupy the country lying to the south of this lake.
''Some time about the feast of all saints, if it be the
will of our good Lord, I intend, with as many of the
french as are willing to encounter the same dangers, to
join the Assiniboels [Assiniboins], who start each year,
just as soon as the streams are frozen over, for the
country of the kaotiouak or Autelsipounes [i. e., *Mandans*], to procure their supply of Indian corn.'' [3] From
the language used by Aulneau it seems clear that these
journeys of the Assiniboins to the Mandans for corn were
a regular part of the yearly round. From Lake Winnipeg to the Mandans: ''the distance is computed to be
two hundred and fifty leagues, but as the party engage
in the hunt as they advance, in all liklihood we shall
go over more than four hundred. . . Toward the
middle of March, I shall leave this place [Mandans] to
return to the shores of ouinipigon.''

Soon after writing this letter, Father Aulneau fell a
victim to a Sioux war party which ambushed and mas-
sacred a detachment of Verendrye's men on an island
in Lake of the Woods. Two years later, however, the
journey to the Mandans which he had contemplated was
carried out by Verendrye.

On October 18, 1738, Verendrye set out with a score
of his Frenchmen and some Assiniboins for the Mandan
villages. The Indians led him a round-about way
through the plains, hunting buffalo as they went, and

3 *Jesuit Relations*, Thwaites, v. lxviii, p. 293.

MANDAN BLUE AND WHITE SPOTTED CORN

finally joined a large camp of Assiniboins. Runners were sent ahead to notify the Mandans of the approach of the Assiniboins and French and to fix a place of meeting, the Assiniboins for some reason (fear of the Sioux?) not caring to take their women and children as far as the villages. On November 28 the place of meeting was reached and the Assiniboins went into camp. That same evening a Mandan chief arrived with thirty men and the four Assiniboin messengers. This Mandan chief was from the most northerly and the smallest of the villages, the "small fort away from the river" with which the Assiniboins appear to have been in closest relations. The chief, however, did not relish the idea of maintaining the large Assiniboin camp which had accompanied Verendrye at free-quarters, for "if they all came to his fort, there must be a great consumption of grain, their custom being to feed liberally all who came among them, selling only what was to be taken away." The chief therefore stated that the Sioux were about to attack his fort and that he was very much pleased that the Assiniboins had come with the French to aid him. The Assiniboins fearing the Sioux greatly, at once determined to remain where they were and send only a picked body of men to the Mandan fort with the French.

The Mandan chief's thirty followers had brought corn in the ear and other articles with them, and now proceeded to trade with the Assiniboins, who had a large supply of English and French goods and also buffalo meat and fat. The Mandans traded "grain, tobacco, peltry and painted plumes, which they know the Assiniboins greatly value" for "muskets, axes, kettles, powder, balls, knives and awls." "They are much craftier in trade than the Assiniboins and others, who are constantly their dupes."

With a score of Frenchmen and 600 Assiniboin warriors, Verendrye marched into the Mandan village on December 3. The Assiniboins at once proceeded to trade, but Verendrye gives no details of the trafficking. He says (p. 19) that the Mandans had besides corn "painted ox-robes, deer skin, dressed buck skin and ornamented furs and feathers, painted feathers, and peltry, wrought garters, circlets for the head, girdles. These people dress leather better than any of the other nations, and work in fur and feathers very tastefully, which the Assiniboins are not capable of doing. They are cunning traders, cheating the Assiniboins of all they may possess, such as muskets, powder, balls, kettles, axes, knives or awls. Seeing the great consumption of food daily by the Assiniboins, and afraid that it would not last long, they set afloat a rumour that the Sioux were near and that several of their hunters had noticed them. The Assiniboins fell into the trap and made up their minds quickly to decamp."

That the Mandans, like all sedentary tribes, were shrewd traders we may well believe, but that they cheated the Assiniboins and other hunting tribes in their exchanges is a statement open to much doubt. The hunting tribes would not have made the long and dangerous trip to the Mandans each year only to be cheated of their goods, and Verendrye himself states that these hunting Indians valued grain more highly than they did European goods, and that the painted robes and feathers, ornamented furs, and other articles made by the Mandans were also highly prized.

Carver (1767) is the next writer to refer to this trade in corn between the Indians of the Upper Missouri and the Assiniboins. He says (p. 109, edition 1781): "On this river [Assiniboin] there is a factory that was built by the French called Fort La Reine, to which the traders from Michillimackinac resort to trade with the Assinipoels and Killistinoes. To this place the Mahahs, who inhabit a country two hundred and fifty miles southwest, come also to trade with them; and bring great quantities of Indian corn to exchange for knives, tomahawks, and other articles. These people are supposed to dwell on some of the branches of the river of the west." These people were evidently the Amahami, a little tribe living with the Hidatsas and Mandans, who were always noted for their daring and for the remarkably long journeys made by their little bands. If Carver's account of this trade may be trusted, it would seem that intertribal wars had stopped the Assiniboin visits to the Mandan villages, and that the Amahamis were taking the corn up through the danger zone to trade it to the Assiniboins and Crees at Fort de la Reine. Carver also states that these two northern tribes were still getting most of their trade-goods from Hudson Bay, just as in Verendrye's day.

This trade in corn with the Assiniboins continued until after the year 1865 and was only ended by the placing of this tribe on reservations. Henry mentions the trade in 1804. Henry (p. 402): "This afternoon the Assiniboins, old Crane and his party, left on their return home to their camp at Moose Mt., all provided with horses, loaded with corn.''

Larpenteur (p. 255): ''At this time a good strong peace had been made with the Assiniboins by the Gros Ventres, Mandans and Rees, and in the fall they would visit and trade corn.''

Boller (p. 122), Assiniboins: ''One old fellow took the lead dragging a broken down bay horse heavily packed with corn, the gift of his Gros Ventres friends.''

Such then was the trade of the Mandans and Hidatsas with the tribes in the plains north of the Missouri during the eighteenth century. At this same period the Mandans, Hidatsas, and Arikaras were engaged in a very similar trade with the tribes in the plains southwest of the river — the hunters of the Black Hills region of western Dakota and of the plains of eastern Wyoming and Montana.

Larocque (1805) met the Rocky Mountain Indians (Crows and some Shoshonis) at the Hidatsa and Mandan villages and accompanied them on their return journey to their own country, west of Powder River and south of the Yellowstone. On p. 22 of his journal [4] he describes the arrival of these Indians at the Upper Missouri villages on June 25, 1805. ''Tuesday 25th. About one in the afternoon the Rock Mountain Indians arrived, they encamped at a little distance from the village with the warriors, to the number of 645, passed through the village on horseback with their shields & other warlike implements, they proceeded to the little

[4] *Journal of Larocque.* Publications of the Canadian Archives, no. 3, Ottawa, 1910.

village, Souliers, and then to the Mandans and returned.
There did not remain 20 person in the village, men, women and children all went to the newly arrived camp
carrying a quantity of Corn raw and cooked which they
traded for Leggins, Robes and dried meat. There are 20
lodges of the snake Indians & about 40 men. The other
bands [Crows] are more numerous."

On the following day the Hidatsa and Mandan warriors visited the camp of the Rocky Mountain Indians
and gave a similar war-parade. The trading then started
in earnest; but Larocque does not give us any account
of this. Farther on, on p. 66, he says: "They have
never had any traders with them, they get their battle
Guns, ammunitions etc from the Mandans & Big Bellys
in exchange for horses, Robes, Leggins & shirts, they
likewise purchase corn, Pumpkins & tobacco from the
Big Bellys as they do not cultivate the ground."

After the Sioux had occupied the Black Hills country
and the Powder River region, the Crows experienced
great difficulty in reaching the villages of the Hidatsas
and Mandans; but they often got through, and their
trade with these villages continued, to some extent, long
after trading-posts had been established in their own
vicinity. Indeed, they did not give up these visits altogether until they were put upon reservations.

Mooney [5] says of the Kiowas: "They have more to
say of the Arikara than of the others, probably because
then, as now, they were the largest of the three tribes,

[5] Mooney's "Calendar History of the Kiowa," in *17th Report*,
Bureau of American Ethnology, part i.

and also, as the Kiowa themselves say, because the Arikara lived nearest.''

As early at least as 1719 the Plains tribes had procured some horses, and they soon began to make raids
and trading expeditions toward the New Mexican frontier. By 1740 they were making regular visits on horseback to the villages on the Missouri, to trade meat, robes,
horses, and Spanish goods for corn, dried pumpkins, tobacco, and English and French goods, particularly Hudson Bay guns and ammunition. The Crow trade at the
Mandan and Hidatsa villages continued after the year
1800, but long before that date the regular visits of the
Kiowas to the Arikara villages were discontinued, owing
to the growing hostility of the Sioux. Thus Trudeau
(pp. 40 and 45) informs us that in July-August, 1795,
the Kiowas and two other Black Hills tribes (apparently
the Prairie Apaches and Arapahoes) were hovering in
the Plains, west of the Arikara villages, not daring to
come nearer. Trudeau also informs us that the Arikaras
had been making raids on the Kiowas — a clear indica-
tion that the Sioux had already engaged in their de-
testable work of stirring up war between peoples that
had lived on terms of friendship for generations. After
being cut off in this manner from the Arikara villages,
the Kiowas continued their trade with that tribe through
the medium of the Cheyennes and Arapahoes. A few
years ago there were several old people among the Southern Cheyennes, since dead, who could remember this
trade. The Cheyennes would take horses, meat, and
robes to the Arikara, Mandan, and Hidatsa villages and
exchange them for corn, dried pumpkins, tobacco, guns,

and European goods. They would then take these articles out to the Black Hills or, in later years, down to the Platte, above its forks, and meet the Kiowas, and perhaps some Comanches, in a trading-fair to which these southern Indians brought large numbers of horses and some Spanish goods, particularly the gaudy striped blankets which the northern Indians so highly prized.

After the Cheyennes moved into the Plains they began to trade at the Arikara and Mandan villages. Some of the Cheyenne bands continued to plant a little corn, tobacco, and perhaps some vegetables until after the year 1800, probably on the lower Cheyenne River, where they often encamped in spring and fall on their return from hunting in the Plains. Perrin du Lac informs us that the Sioux made a practice of robbing the Cheyenne corn and tobacco patches while that tribe was away hunting in the Plains. The Cheyennes therefore could not depend much on their own crops, even as a partial supply, and made regular journeys to the Arikara and Mandan villages to trade for corn, pumpkins, and tobacco. The Sioux, playing their usual part as trouble-makers, frequently stirred up the Mandans or Arikaras against the Cheyennes, or the Cheyennes against these tribes, and broke up the trade. But if the Cheyennes were at war with the Mandans they were probably at peace with the Arikaras, and in spite of all the bickering almost every spring and autumn found large camps of Cheyennes coming in to trade at one of the villages on the Missouri.

The Sioux, though among the best customers of the agricultural tribes, were never really their friends and could never be trusted. They were the Picts of the Up-

per Missouri, continually harassing the village dwellers.
It is probable that they procured as much corn by plunder and extortion as in honest trade. The western Sioux,
those of the Missouri, occasionally attempted to settle
down to an agricultural life, but they never persisted for
any length of time.[6]

The Arikaras were probably most under the influence
of the Sioux and suffered most from plundering and in-
timidation. However, they appear to have reaped some
benefits from this very trying intimacy. Trudeau is the
first to give some account of the relations existing between these two tribes. In his journal, 1795, he says
(p. 47): "The Ricaras and this Sioux Nation [the 'Ta
Corpa' band of Sioux[7]] live together peacefully. The
former receive them in order to obtain guns, clothes, hats,
kettles, cloths, etc., which are given them in exchange
for horses. They humor them through fear and to avoid
making too many enemies among the Sioux, who would
inevitably overpower them.

"The last frequent the Ricaras and make them great
promises to live with them in peace and harmony, in
order that they may smoke their tobacco, eat their In-

---

[6] The Minniconjou Sioux, according to the information given
by some members of the band, secured seed from the Arikaras
almost sixty years ago and have raised this corn ever since. The
Yanktons appear to have raised some corn on the east bank of
the Missouri, in the vicinity of the town that now bears their
name, as early at least as the year 1850. They were always
more inclined to an agricultural life than the other Missouri
River Sioux, probably owing to their closer connection with the
Santees and other eastern bands that practiced agriculture.

[7] "Tarcoehparh" — given by Lewis and Clark as one of the
three bands of the Minniconjou Sioux.

dian wheat and hunt freely the wild oxen and beaver on
their lands during the fall and winter. In spring they
withdraw to the other shore [*i. e.*, the east bank of the
Missouri], from whence they usually return to steal
their horses and sometimes to kill them.''

Perrin du Lac gives a very similar account of the relations of the Sioux with the Arikaras and other tribes
with which they were on so-called friendly terms.

Lewis and Clark (p. 144) Arikaras: ''They claim
no land except that on which their village stands, and
the fields which they cultivate. Though they are the
oldest inhabitants, they may properly be considered the
farmers or tenants at will of that lawless, savage, and
rapacious race, the Sioux Teton, who rob them of their
horses, plunder their gardens and fields and sometimes
murder them without opposition. . . They maintain
a partial trade with their oppressors, the Tetons, to
whom they barter horses, mules, corn, beans and a species
of tobacco which they cultivate.'' And, p. 165: The
Arikaras obtain peltries ''not only by their own hunting, but in exchange for corn from their less civilized
neighbors . . . being under the influence of the
Sioux, who exchange the goods they get from the British
for Ricara corn.''

Henry (p. 339) states that when the Sioux came to
attack the Hidatsas they ''compelled the Mandans to
provide them with corn, beans, squashes, etc., for their
sustenance.''

Mrs. Holley [8] tells us that in 1843 the Mandans and
Arikaras left Primeau in charge of their abandoned

------

[8] *Once their Home, or Our Legacy from the Dakotas*, p. 179.

summer villages and corn caches. The Sioux came, and
in spite of Primeau's protests and warnings, broke into
the caches and robbed them.

From the time when the Sioux first made their appearance on the Missouri the village Indians appear to
have attempted from time to time to combine against
them. Thus Lewis and Clark tell us that in 1804 the
Arikaras sent embassadors to the Mandans to arrange
peace. On their way home the embassadors were way-
laided and whipped with quirts by the Sioux. The
Sioux then induced a number of young Arikaras to join
them in a raid on the Mandans, and thus set the two
tribes to fighting again. The same authors mention
another trick used by the Sioux in stirring up trouble
between these village tribes. The Sioux made a raid on
the Mandans and scattered some Arikara corn on the
ground, "to induce a belief that they were Ricaras."
Even as late as 1867 we find an Indian agent reporting:
"The Arickarees, Gros Ventres and Mandans are at Ft.
Berthold in a truly pitiable condition. . . They are
hemmed in by all bands of Sioux: by those we call
friendly as well as by the hostile bands. . . The
Sioux have killed a number of them this spring, and are
very vindictive toward them." [9]

Hayden (p. 352): "The second market for their
grain is with several bands of Dakotas, who are at peace
with them. These Indians make their annual visit to
the Arikaras, bringing buffalo robes, skins, meat, etc.,
which they exchange for corn."

In the fall of 1853, 2500 Sioux came to the Arikaras

[9] *Report* of the Commissioner of Indian Affairs for 1867, p. 346.

to trade meat and robes for corn, dried squashes, and beans, and to steal whatever they could lay hands upon. They then "left for the buffalo country, taking care to set the prairies on fire in order to prevent the buffalo from visiting the Ree country — an act of dastardly malignancy, as it deprived the Arickarees of the means of support for their horses and cattle." [10]

Boller speaks of the distrust the village tribes felt for even the "friendly" bands of Sioux (p. 159): "They expected . . . tidings of the whereabouts of the Blackfoot and Uncpapa bands of Sioux, who being just now on friendly terms were likely, since the corn was gathered in, to visit the Rees and Gros Ventres to maintain the entente cordiale." This author also describes the straits to which the Sioux were often reduced during a hard winter and how on such occasions they sought peace with the agricultural tribes. Boller, p. 209: "Many were obliged to kill their horses to avoid starvation, and there were rumors of the Medicine Bear's band desiring to make peace with the Rees and Gros Ventres in order to procure corn." And again (p. 210): "Small party of Yanctowahs, numbering about sixteen lodges, came to the Gros Ventres to make peace and relieve their pressing necessities by trading corn." Finally (p. 262): "The Sioux expressed themselves very anxious to make peace with the Rees and Gros Ventres as they invariably do when starving, in order to trade corn, but the latter placed no confidence in their Punic faith."

In early times the Mandans and their agricultural

[10] *Pacific Railroad Surveys*, v. i, p. 265.

neighbors sometimes went out into the Plains to meet the hunting tribes and trade with them. In 1804 Henry accompanied such a party of Mandans and Hidatsas out into the plains to meet the Cheyennes and Arapahoes. The party consisted of a large force of armed warriors and many women, some with their children, bringing loads of corn, beans, and dried squashes to trade to the Cheyennes. The women, says Henry, "had their horses loaded with corn, beans, etc., themselves and children astraddle over all, like farmers going to the mill." And after reaching the Cheyenne camp, "the women were also busy exchanging their corn for leather, robes, smocks — as if at a country fair."

It will be noted that all of this trade in corn, beans, squashes, and European goods with the hunting tribes of the plains came to the villages of the Mandans, Hidatsas, and Arikaras. The agricultural tribes farther south, in Nebraska — Ponkas, Omahas, Otoes, and Pawnees — do not appear to have had any regular trade with the Plains Indians in early years, either in agricultural products or European goods. One reason for this condition of affairs was that the village tribes of the Nebraska region were constantly at war with the tribes in the Plains. Another reason was that, up to the year 1820, these village Indians rarely had sufficient European goods to supply their own needs and, setting aside their usual attitude of bitter hostility toward the Nebraska tribes, the Plains Indians preferred to take the longer journey to the Arikara or Mandan villages, where they could procure not only corn and vegetables but also guns, ammunition, kettles, and whatever else they had need of.

The Omahas had an irregular trade in corn with their kinsmen the Ponkas, who sometimes planted fields of their own, but more often preferred to trade robes and meat to the Omahas for what corn and vegetables they required. Dunbar states that the Pawnees in early times traded corn to the Mandans: a surprising assertion which may have been based on some vague tradition that the Pawnees supplied seed to the Mandans when that tribe first arrived on the Missouri. There is certainly no mention of such a trade during the historical period in any of our sources of information.

2. Trade with the Whites

Besides their trade with the tribes in the Plains, the agricultural tribes of the Upper Missouri early developed a trade in corn and vegetables with the white traders and explorers. To the Indians this trade was certainly a great benefit, as it enabled them to procure much-coveted articles of European make; on the other hand it seems highly probable that the white explorers and traders would have found it impossible to carry on their operations without the supplies of food obtained from the village Indians. That this trade was established with the first coming of the whites and that it grew to large proportions may be easily demonstrated.

Like the non-agricultural tribes of Indians, the whites who lived in the Indian country always had a craving for vegetable food, especially corn, and never overlooked an opportunity for obtaining a supply. They always took with them as much corn as they could conveniently carry on their journey into the Indian country. Mackinac Island, in the straights between Lake Huron and Lake Michigan, was a famous corn market of the early fur-traders. The great town of the Sacs on the Fox-Wisconsin River route to the Mississippi, was another center of this trade. Carver (p. 24) says: "In their plantations which lie adjacent to their houses and which are neatly laid out, they raise great quantities of Indian corn, beans, melons, etc. For this place is esteemed the best market for traders to furnish themselves with provisions of any kind within 800 miles." Prairie du Chien on the Upper Mississippi was another center of the corn trade in early times; while on the Upper Missouri the villages of the Mandans, Hidatsas, Arikaras, and Omahas were the corn markets for parties going into the plains or to the Rocky Mountains.

As we have already seen, Verendrye in 1738-39 procured large quantities of corn and meal from the Mandans. Trudeau (1795) is the next to mention this trade definitely. He says (p. 27) that the prices demanded by the Arikaras for their corn and other provisions had been greatly increased through the foolish action of several French traders who came among these Indians in 1793 and "paid the Indians three prices for food, a habit and rule which took such deep root in the minds of these people that had I not taken the precaution to provide myself with dried meats from the Cheyennes and Sioux, who sell them more cheaply, I should have run great risk of fasting."

Henry also speaks of this trade (p. 328): "We purchased sweet corn, beans, meal and various other trifles. Having bought all we required, which was 3 horse loads,

we were plagued by the women and girls who continued
to bring bags and dishes full of different kinds of
produce.''

Bradbury (1811) mentions the trade with the Omahas:
''We found a considerable number of the Indians as-
sembled to trade. They gave jerked buffalo meat, tallow,
corn and marrow'' (p. 87).

The Oto missionary, Merrill, states in his diary that
in 1834 a trader named Edwards told him that the
Omahas traded corn to the fur-traders at the rate of one
bushel of corn for one yard of calico, the trade-price of
this cloth being one dollar per yard (p. 173).

Lewis and Clark were perhaps the heaviest purchasers
of corn of all the whites who visited the Mandans and
Hidatsas in early years. The good fortune of these explorers fixed their winter-camp at the villages of these
agricultural tribes, a location which certainly prevented
much suffering, if not actual death, among their men.
The following series of short extracts from the journals
of the expedition will give a good idea as to the extent
of their purchases from day to day. Some of their
larger acquisitions of corn, running up into hundreds
of bushels, have already been referred to.

Lewis and Clark (Nov. 16): ''An Indian came
down with 4 buffalo robes and some corn, which he
offered for a pistol, but was refused.'' Mandans (Dec.
21): ''A woman brought her child with an abscess in
the lower part of the back, and offered as much corn as
she could carry for some medicine. . .'' (Dec. 22):
''A number of squaws and men dressed like squaws
brought corn to trade for small articles with the men.''

(Jan. 5): "The Indians continue to purchase repairs [*i. e.*, pay for blacksmith work] with grain of different kinds." (Jan. 16): Kagohami visited us and brought us a little corn." (Jan. 20): "A number of Indians visited us with corn to exchange." (Feb. 5): "A number of the Indians came with corn for the black-smith . . . who now being provided with coal has become one of our greatest resources for procuring grain. . . The blacksmith cut up an old cambouse, of metal, we obtained for each piece of 4 inches square, 7 or 8 gallons of corn from the Indians who were de-lighted with the exchange." (Feb. 18): "Our stock of meat is exhausted, so that we must confine ourselves to a vegetable diet. . . For this, however, we are at no loss, since both on this and the following day our blacksmith got large quantities of corn from the In-dians . . ." (March 13): "Many Indians, who are so anxious for battle axes that our smiths have not a moment's leisure, and produce us an abundance of corn."

Henry (p. 325) mentions the eagerness of the Indians to trade their corn, etc.: "They soon asked us to trade and brought buffalo robes, corn, beans, dried squashes, etc."

The following extracts refer to the early trade with the Arikaras.

Manuel Lisa at the Arikara villages: "The women then appeared with bags of corn with which to open trade but an Indian rushed forward and cut the bags with his knife." [11]

11 *American Fur Trade*, v. i, p. 117.

The winter spent by Maximilian at Fort Clark, among
the Mandans, would certainly have been a disastrous one
for the whites had it not been for the supply of corn
which the post-trader had secured from the Indians.
This corn was for much of the time the only barrier between the men at the fort and starvation. Maximilian
(p. 36): "A high stage was stretched in the court yard
where a part of the stock of maize was deposited. . .
We were forced to live on hard dried meat and boiled
maize." The engages at the fort: "The poor fellows
had no meat and had lived on maize boiled in water"
(p. 48). "We had nothing but maize and beans and
the water from the river" (p. 58). "We have lived on
nothing but maize boiled in water, and this was really
the case with many persons at this place" (p. 61). "We
subsisted entirely on maize broth and maize bread" (p.
63). "As our supply of beans was very low our diet
consisted almost exclusively of maize boiled in water"
(p. 76).

That this dependence of the fur-traders upon Indian
corn was not confined to the posts in the immediate
neighborhood of the agricultural villages, but that large
quantities were shipped to the more distant posts on the
Missouri and Yellowstone is evident from the references
we find to such shipments.

Maximilian (p. 82): "On the 15th of April Picotte
arrived with about twenty men and had his boat laden
with maize which he was to carry to Fort Union."
Maximilian (p. 211): "A boat laden with maize belonging to Mr. Campbell here passed us, it had left the
Mandan villages a fortnight before."

We have an account of this corn trade in Hayden, in considerable detail; and Matthews also mentions it. In the *American Fur Trade* there is a statement which shows the extent of the trade and the dependence of the traders on corn.

Hayden (p. 352): "Whatever is concealed [cached] in this way is intended to remain in the ground until the succeeding spring, at which time, the buffalo usually being far distant, it is their only source of food. Besides the great advantage acruing to themselves over other wandering tribes, by tilling the soil, they have two markets for their surplus produce. The first is the fort of the American Fur Co. located near their village, at which they trade from 500 to 800 bushels in a season. This trade on the part of the Indians is carried on by the women, who bring the corn by panfuls and the squash in strings and receive in exchange knives, hoes, combs, beads, paint etc., also tobacco, ammunition and other useful articles for their husbands. In this way each family is supplied with all the smaller articles needed for a comfortable existence; and though the women perform all the labor, they are compensated by having their full share of the profits."

"The harvest generally took place in October. The corn fields were unfenced and were frequent objects of raids by hostile tribes. The traders made extensive use of the maize and all the larger posts had mills to grind it." [12]

"In favorable years they had good harvests and were able to supply other Indians and to their traders be-

---

[12] *American Fur Trade*, p. 807.

sides keeping all they wanted for their own use'' (Matthews, p. 12).

An episode in the early fur-trade which illustrates the quantity of corn the whites were able to obtain from the Mandans and their neighbors, was the establishment of a whiskey still at Fort Union, at the mouth of the Yellowstone, at the time when the government's agents on the Lower Missouri were very active, making it very dangerous for the American Fur Company to attempt to smuggle liquor up the river in its steamboats.

American Fur Trade (p. 358) quotes a report from Fort Union describing this experiment: ''Our manufacture flourishes admirably. We only want corn to keep us going. The Mandan corn yields badly but makes a fine sweet liquor.'' Wyeth, a rival trader, gave information to the government agents and an investigation was ordered. The men at Fort Union coolly explained that the still had been set up for the purpose of carrying on scientific experiments to determine whether a good wine might not be made from wild fruits which abounded on the Upper Missouri; and their explanation was accepted by the government; but the fur company was so badly frightened that it ordered the whiskey still abandoned.

VI—THE SACRED CHARACTER OF CORN

1. THE CORN AND THE BUFFALO. 2. CORN ORIGIN MYTHS

Primitive peoples have always been strongly drawn toward the worship of their principal sources of food supply, and among the Indians of North America this tendency has been perhaps one of the most marked characteristics of the religious thought of the race. The hunting tribes looked upon the game animals on which they subsisted as sacred, many of their religious rites were intended to procure the aid of these animals; and as the people considered themselves closely related to the animals that gave them life, they often organized themselves into clans or gentes named after these animals. Thus the Elk Gens would be kindred of the elk, would bear the elk totem, practice elk rites, and have a tabu on the eating of certain parts of this animal. In the same way, among the sedentary agricultural tribes the plants cultivated were looked upon as sacred. This was particularly true of the corn, always the main resource, and often considered the mother of the whole tribe or of certain clans or gentes within the tribe. Corn names for men and women, indicating a close kinship of the people with the corn, are found in all of the agricultural tribes of the Upper Missouri area, as also among the Sioux; corn tabus are found in at least three of these tribes; but we do not find any trace of the corn clans or gentes, which were a common feature of the social organization of the more advanced agricultural tribes in the South-west and on the Lower Mississippi.[1]

The village Indians of the Upper Missouri depended largely on the buffalo as a means of support and con-sidered that animal sacred; but the corn also played an important part in the lives of the people, and we find evidence that at least certain divisions in some of the tribes looked upon the corn as more sacred than the buffalo. We also find old corn rites changed into buffalo rites, indicating that at an earlier period the corn held the higher position in the people's estimation. Corn has a prominent place in the creation traditions, myths, and tales of all of the agricultural tribes of this region; and although many of the old customs and beliefs have been long since discontinued and forgotten we still find quite an extensive body of corn rites, beliefs, and practices.

The Pawnees, between whose culture and that of the Pueblos of the Southwest some similarity exists, are said to have had the largest volume of ceremonial observances, connected with the corn and its cultivation, of any tribe in this area. As the Arikaras are really a branch of the Pawnees, we would naturally expect them to have corn rites similar to those of the Pawnees; but we find that the beliefs and practices of the two peoples have very little in common. This is probably due to the fact that the Arikaras have been living among Siouan tribes for

[1] The large number of corn clans among the various Pueblo tribes is particularly noticeable. The Pueblos of Acoma had no less than five of these: the Red, Yellow, Blue, Brown, and White corn clans. These clans were grouped together into a fratry known as the Corn People.

at least two hundred years, and that their religious concepts, their rites, and their mode of life have been profoundly affected by their long intimacy with these neighbors of alien stock. Of the Siouan tribes of this region, the Mandans and the Omahas appear to have had the most highly developed body of rituals and beliefs relating to the corn and its cultivation.

The religion of the Pawnees was a star cult overlaid with a very highly developed system of ritualistic practices. Their name for the Creator was Tirawa. His wife was the goddess Atira, a name meaning literally "Comes-from-Corn" or "Born-from-Corn." In the beginning Tirawa tells the Evening Star — Bright Star — to stand in the west. She is to be "the Mother of all things." Great Star — the Morning Star — is to stand in the east. He is to be a warrior. Bright Star has a garden in the western skies in which all things to be placed on earth are created: all animals and birds, and even the sun's fire (light).[2] After the earth is made by Tirawa himself, all of these things are placed upon the earth by the four servants of Bright Star: Wind, Cloud, Lightning, and Thunder. Great Star comes from his position in the east and marries Bright Star in her garden, and from this marriage a girl is born in winter. The Sun marries the Moon and they have a son born in summer. These first two human beings are now placed upon earth, where they marry. Other stars create other human beings and place them on earth. Each pair, man

---

[2] The altar which was erected in every Pawnee lodge was given the same name as this garden of the Evening Star in the west: *wiharu*, "the place of the wonderful things."

MANDAN SOFT WHITE CORN

and wife, is given a sacred bundle by the star that created them, and in this bundle is always an ear of sacred corn. In one of the versions of this creation myth we are told that all of these peoples spoke the same language; that the people who lived in the southwest had white sacred corn, those in the northwest yellow, those in the northeast black, and those in the southeast red.[3] It will be noted that Tirawa, the Creator, was far away in the heavens and had no direct intercourse with the people on earth. The Pawnees looked upon him with awe and dread. Their dealings with him were mostly carried on through the medium of the beloved Bright Star who stood in the west and who was represented on earth, symbolically at least, by the Mother Corn — the sacred white corn, which the people kept in their sacred bundles, on the altars in the lodges, and which they brought into so many of their rites.

Although the Arikara traditions of the present day are very different from those of the Pawnees, we find among them some traces of this older star cult. Thus we are told in some of the Arikara creation traditions that people were created by gods, or stars, and were placed upon earth; that a god or star had a garden in the sky, and that from this garden Mother Corn was sent to the earth to help the people. The Arikaras certainly

---

[3] Dorsey, *Traditions of the Skidi Pawnee*, p. 10. White was the color of the Evening Star and was female; red was the color of Great Star, the Morning Star, and was male; black was the color of Black Star, who stood in the northeast, who had the power to send animal gods to help the people; and yellow was the color of Yellow Star in the northwest, who had the power to send buffalo to the people.

reverenced the Mother Corn as much as did the Pawnees, but they appear to have recognized her as the First Mother, while the Pawnees more often looked upon the Mother Corn as the earthly representative of the First Mother, who was Bright Star.

Among the Siouan tribes of this region we find no trace of a star cult such as that of the Pawnees, and their views as to things as they were in the beginning are very unlike those of the Pawnees. Their stories of the creation usually open with the statement that there was a being (Wakonda — God) floating on the waters; that he created the earth from a pinch of mud brought up to him by diving water-birds, and that he then created people and animals and placed them on the earth. We find no clear trace of a belief in the heavenly origin of the people or of corn, as among the Pawnees; but the Siouan tribes often spoke of corn as a "mother" and there is some reason for believing that they looked upon the corn, as the Pawnees also sometimes did, as symbolizing the Mother Earth and her fruitfulness.

Speaking of the great reverence the Mandans had for the corn, Scattered Corn stated that corn was always treated with the greatest respect and care; no grains were ever left scattered about and the stalks were never touched with metal knives.[4] The empty corn caches were purified and blessed before the corn was placed in them.

The Omahas considered the corn a "mother," and one of their myths describes the birth of the corn, as well as

---

[4] Because metal belonged to or was related to Thunder, which was a war or destructive power? We find this belief in the sinister influence of metal among many tribes.

the buffalo, from Mother Earth.[5] The great reverence which this people had for the corn is exhibited in the following quotation from their sacred legend:

"The maize being one of the greatest means to give us life, in honor of it we sing. We sing even of the growth of its roots, of its clinging to the earth, of its shooting forth from the ground, of its springing from joint to joint, of its sending forth the ear, of its putting a cover on its head, of its ornamenting its head with a feather,[6] of its invitation to men to come and feel of it, to open and see its fruit, of its invitation to man to taste of the fruit."

Many of these old Omaha songs that had to do with every phase of the planting and cultivation of the corn are now lost.

1. The corn and the buffalo

Among most of the agricultural tribes of the Upper Missouri we find traces of two antagonistic forms of culture, one of which gives to the corn the higher position in the sacred traditions and rites, while the other puts the buffalo first and relegates the corn to a subordinate

---

[5] *27th Report,* Bureau of American Ethnology, p. 147. This is the well known tale of the old woman who lived in a cave underground and kept the corn, buffalo, and other animals. Their coming forth from the cave symbolizes their birth from Mother Earth.

[6] By the "feather" is evidently meant the central upright spike of the corn tassel. The feather head-ornament peculiar to the Omahas with the long central upright feather and the smaller drooping feathers surrounding it, resembles a corn tassel. The Pawnees liken the tassel of the sacred white corn to an eagle feather.

place.   The tribes that had altars in their lodges — the
Pawnees, Arikaras, Mandans, and Hidatsas — placed
both an ear of Mother Corn and a buffalo skull on the
altar.

Among the Pawnees, according to Mr. Dorsey, the corn
held a higher place than the buffalo; they brought in the
corn in all of their ceremonies, from the great Hako
ritual down to the little household rites and practices of
daily life.[7]   But when we examine the sacred traditions
of the people, the right of the corn to the higher position
is less clear.   The traditions seem to indicate that among
the Pawnees as among all of the other Upper Missouri
village tribes, part of the people regarded the corn as of
higher importance, while others gave that position to the
buffalo; but both the tales and the sacred rites and household customs seem to indicate that the reverence for the
Mother Corn was older and much more deeply rooted in
the people's thoughts than was the reverence for the
buffalo.

Among the Arikaras we seem to find evidence that the
corn was regarded with greater reverence than was the
buffalo.   The leading rôle played by the star gods and
goddesses in the Pawnee creation traditions is given to
Mother Corn by the Arikaras.   She leads the people out
of the ground after the great flood and guides them to
their historic home in the Missouri Valley.[8]   "In the
series of rites, which began in the early spring when the
thunder first sounded, corn held a prominent place.   The
ear was used as an emblem and was addressed as

---

[7] Dorsey, *Traditions of the Skidi Pawnee*, p. xv.

[8] Dorsey, *Traditions of the Arikara.*

'Mother.' Some of these ceremonial ears of corn had been preserved for generations and were treasured with reverent care. Offerings were made, rituals sung, and feasts held when the ceremonies took place. Rites were observed when the maize was planted, at certain stages of its growth, and when it was harvested. Ceremonially associated with maize were other sacred objects, which were kept in a special case or shrine. Among these were the skins of certain birds of cosmic significance, also seven gourd rattles that marked the movements of the seasons.'' [9]

As among the Pawnees, we find among the Arikaras some traditions that give to the buffalo the leading rôle; but the tales and rites in which the Mother Corn comes first are far the more numerous.

Among the Mandans the ceremonies and traditions relating to corn seem to be fully as important and as ancient as those that have to do with the buffalo. The most interesting feature of the traditional accounts of the creation and early history of the Mandans is a dual set of tales. One set gives the corn the higher place, and in some respects these tales resemble those of the Arikaras, though in other respects they differ widely. The other group of tales has very little to say of the corn, and these stories correspond rather closely to some of those of the non-agricultural Siouan and Algonquian neighbors of the Mandans. The Mandan traditions and rites are so interwoven with those of the Hidatsas that it is almost impossible to separate these belonging to one tribe from those belonging to the other. There are two possible explana-

---

[9] *Bulletin 30*, Bureau of American Ethnology, v. i, p. 85.

tions of this dual set of Mandan-Hidatsa ceremonies and traditions.

The first theory is that the non-agricultural series is the one the Mandans brought with them into the Missouri Valley, and that the second series was acquired from the Pawnees and Arikaras from whom, according to this theory, they first procured corn and the rites and beliefs which accompanied the practice of agriculture. This theory has much to recommend it; but it is stoutly denied by the Mandans themselves.

The second explanation is that the agricultural series of traditions and rites belonged to the old Mandan tribe, and that the non-agricultural set has been acquired from the Hidatsas during the long period of close association with that tribe. This explanation also has its supporters. Its strongest points are that the Hidatsas claim that all of their agricultural knowledge came from contact with the Mandans, and that the Mandans cannot have come through the terrible experiences which, on more than one occasion, threatened them with complete extinction as a tribe, without losing a great part of their older traditions and rites.

There is also the third possibility that the Mandan traditions and rites of today are the result of the fusing of three sets: Mandan, Hidatsa, and Arikara. There are arguments that may be used in support of any of these three theories, but it seems impossible to prove any one of them or to learn whether in early years the Mandans did or did not reverence the corn more highly than the buffalo.

Among the Omahas, however, we have clear evidence

that the old tribal organization was on an agricultural basis and that the corn rites, many of them long since discarded and now almost forgotten, were much older than the rites connected with the buffalo.[10] Here we need pay no heed to the usual tales which recount how seed was first obtained from the Arikaras, or Pawnees; for the evidence that the Omahas had practiced agriculture for generations before they came in contact with these Caddoan tribes is overwhelming.

To sum up: it appears that we may safely judge that at one time the Pawnees, Arikaras, Omahas, and Ponkas[11] held the corn to be more sacred than the buffalo, and that it also appears clear that the corn cult among these tribes was much more highly developed in early times, before the Indians secured horses and European weapons, the rise of the buffalo cult in most cases being a later development. In the case of the Mandans, Hidatsas, Otoes, and Iowas, the old traditions and rites on which we must base our judgment have been lost or have become so corrupted that it is almost impossible to arrive at any safe conclusion.[12]

---

[10] This evidence will be quoted in Chapter VII under the heading, Ceremonial Organization.

[11] The Ponkas were a part of the old Omaha tribe, or rather, the Ponkas and Omahas were one tribe, until the time of their arrival on the Missouri.

[12] The Otoes and Iowas are said to have been very closely connected with the Omahas and Ponkas during the migration to the Missouri. We might therefore conjecture that their culture was similar to that of their Omaha-Ponka kindred. The Iowas were certainly good agriculturists before the year 1700, and they still have two varieties of sacred corn at the present day and corn tabus.

2. Corn origin myths

Most of the agricultural tribes of this area still preserve myths or traditions which explain in detail the
sacred origin of corn, and several of these tribes have two
or more very divergent accounts of the first coming of
the corn — a fact that is not easily explained unless we
suppose that they have borrowed from neighboring tribes
or that, as in the case of the Arikaras, the tribe as it
exists today is composed of the fragments of two or more
formerly independent tribes or groups, each of which
had its own sacred tradition of the origin of corn.

Among the Caddoans — the Pawnees, Arikaras, and
Wichitas — we find some faint vestiges of a belief that
the corn originated in the heavens and was sent to the
earth by the gods or goddesses.

The Pawnees considered Bright Star, the Evening
Star, the mother of all things, and believed that in her
garden in the western skies the corn was always ripening.
Arisa, the leading Skidi Pawnee priest, who died very
old in 1878, told the following myth of the origin of
corn: [13] Tirawa created or caused to be created a boy
and a girl and placed them on earth; they married and
had children. Their son followed a meadow lark, the
messenger of the four servants of Bright Star: Cloud,
Wind, Thunder, and Lightning. The bird led the boy
far away and finally to an earth-lodge in which these
four servants were sitting. They taught the boy how to
live; they gave him the buffalo to kill and a sacred
bundle containing seed; they taught him how to make
hoes with the shoulder blade of the buffalo and gave

[13] Dorsey, *Traditions of the Skidi Pawnee*, p. 21.

him sacred rites. This boy returned to the people and
became the head-priest; he kept the sacred bundle of
seed, giving the seed to the people in the spring and
receiving fresh seed from them again in the autumn.
This tale implies that corn was given to the people by
Bright Star and came from her garden in the sky. The
Arikara have a similar creation tradition which states
that a god (or goddess) had a garden in the sky in
which corn was growing, and that it was from this garden that Mother Corn came to the earth to help the
people. In another Skidi myth [14] we are informed that
Tirawa gave seeds of all kinds to Spider Woman, who
lived under the earth, instructing her to grow crops and
distribute the increase among the people so that all might
have seed. Another version states that the Sun and
Moon had given the seeds to this old woman.[15]

The Arikara creation traditions, collected by Mr. Dorsey, state that the gods made people. Some were giants,
some little people. The giants grew insolent and wicked
and the gods decided to send a great flood and destroy
them; but before sending the flood the gods made certain animals and sent them to save the little people, who
were good. These little people were changed into grains
of corn and were taken under the ground by the various
animals. Some of the tales state that after the flood
had subsided these animals led the people out of the
ground again and guided them westward to their historic
home in the Missouri Valley, but other versions declare
that it was Mother Corn who brought the people out of

---

[14] Dorsey, *Traditions of the Skidi Pawnee*, p. 39.

[15] Same, p. 36.

the ground and took them west. There are at least ten versions of this story among the Arikaras, all agreeing on the main points but differing in details. One story gives the mouse the credit for taking the people into the ground and later leading them forth again; another says the fox was the leader, a third the badger, a fourth that it was a group of animals, each doing his share, while the other versions state that the animals led the people under the ground and that Mother Corn assisted by the animals led them forth again.

Mr. Dorsey's third tale [16] gives a very prominent place to Mother Corn. This version opens with the creation of people by Nesaru (God). The large people (giants) thinking themselves as powerful as Nesaru grow insolent. Nesaru turns the good people into grains of corn and has the animals take them under ground, into a great cave; he then sends the flood and drowns the wicked giants. "Nesaru in the heavens planted corn in the heavens, to remind him that his people were put under ground. As soon as the corn in the heavens had matured, Nesaru took from the field an ear of corn. This corn he turned into a woman and Nesaru said, 'You must go down to the earth and bring my people from the earth.' She went down to the earth and she roamed over the land for many, many years, not knowing where to find the people. At last the thunders sounded in the east. She followed the sound, and she found the people underground in the east. By the power of Nesaru himself this woman was taken under ground, and when the people and the animals saw her they rejoiced. They

[16] Dorsey, *Traditions of the Arikara*, p. 12.

OMAHA, IOWA, AND OTO CORN

1. Omaha white	4. Omaha black	7. Iowa blue
2. Omaha blue	5. Iowa sacred red	8. Oto white
3. Omaha brown	6. Iowa sacred brown	9. Oto black

knew her, for she was the Mother-Corn.   The people and
the animals also knew that she had the consent of all the
gods to take them out.''

Mother Corn now calls for aid, and the mole, the
badger, and the long-nosed mouse come forward and begin to dig a passage from the cave to the surface of the
earth. Mother Corn starts to lead the animals and
people forth, but as she reaches the mouth of the tunnel
the thunders sound in the east, the earth heaves and the
animals and people are cast out upon the surface of the
earth.   Some of the people wish to remain at this place,
so they are changed into moles, badgers, and mice and
live with these animals in holes in the ground.   Mother
Corn leads the rest of the people toward the west.   They
have many adventures before crossing the Missouri.
Mother Corn now returns to heaven and comes back with
a man.   Some of the tales state this was Nesaru himself.
He establishes the office of chief and gives the people
rules of life.   Mother Corn says to the people: ''The
gods in the heavens are the four world-quarters, for they
are jealous.   If you forget to give smoke to them they
will . . . send storms.   Give smoke to me last.   The
Cedar-Tree that shall stand in front of your lodge shall
be myself.[17] I shall turn into a Cedar-Tree, to remind
you that I am Mother-Corn, who gave you your life.   It
was I, Mother-Corn, who brought you from the east.   I

---

[17] Among the Omahas also the cedar represents life and is
connected with thunder. The Omahas had a sacred cedar pole,
which belonged to the gens that kept the thunder rites. This
gens also had the sacred shell, and it seems probable that the
pole, and perhaps the shell also, was formerly connected with the
cultivation of the corn. See Fletcher and La Flesche, p. 194.

must become a Cedar-Tree to be with you. The stone that is placed at the right of the Cedar-Tree is the man who came and gave you order and established the office of chief. It is Nesaru . . . watching over you. It will keep you together and give you long life.'' [18]

Among the Pawnees there appears to be no trace of this story of the flood and of Mother Corn leading the people out of the ground, but the Wichjtas have a tradition that during the flood a woman took people and seeds upon a raft and saved them.[19]

The Arikara traditions agree that after Mother Corn had led the people into the Missouri Valley she turned herself into corn and thus provided seed for the people to plant. In one of the tales the whirlwind becomes angry at Mother Corn and comes to attack the people; Mother Corn turns herself into a cedar tree and a rock (the black meteoric star of the eastern heavens, representing Nesaru?) falls beside the cedar tree. Whirlwind strikes the cedar tree (Mother Corn) injuring her, and she throws out first an ear of red corn, then a yellow, then a black, last a white ear (the four sacred colors). This Cedar Tree becomes the ''Wonderful Grandmother'' and is placed before the Arikara medicine-lodge, while the black meteoric rock becomes the ''Wonderful Grandfather'' and is placed beside the Cedar Tree. Mother Corn then gives the four sacred ears to the people: ''My people, this corn is for you. They are seeds. You shall plant them, so that in time you can offer this corn to the gods also. This will be done to

[18] Dorsey, *Traditions of the Arikara*, p. 17.
[19] Dorsey, *Traditions of the Skidi Pawnee, footnote* 74.

remind them that I was once Corn up in the heavens and was sent down to take you from the ground.'' [20] Mother Corn taught the people the rites associated with the sacred bundles; she then returned to the place in the east, from which she had come. Before departing she bids the people bring her all of the children's moccasins which she ties in a bundle, placing the bundle on her back. She then turns herself into an ear of corn and the people throw her, and the children's moccasins, wrapped up in a buffalo skin, into the river, so that she may return to her place in the east. After many years she returned one autumn and taught the priests more bundle ceremonies and songs. Then she departed and was never seen again.[21]

The Pawnees do not appear to have had any traditions telling of Mother Corn leading the people out of the ground, but they had tales of the Spider Woman who kept the buffalo and corn in a cave underground. This old woman lived in the center of the earth in a vast cave in the far north, where the sky touches the earth. Tirawa gives her seeds of all kinds and bids her plant the seeds and give the increase to the people for seed; but she hoards her crops and gives the people nothing. Tirawa sends the Sun Boys against the old woman; they cause the grasshoppers to attack her and carry her up and put her in the moon, where she may still be seen. The Sun Boys then give all of her hoarded seeds to the people.[22] Another version says that the Spider Woman

[20] Dorsey, *Traditions of the Arikara*, p. 22.

[21] Same, p. 22.

[22] Dorsey, *Traditions of the Skidi Pawnee*, p. 39.

kept all of the seed and also spun a great web to keep all of the buffalo underground. The buffalo ask her to permit them to go south to the people; she refuses; but they attack her and trample her into the ground. They then go south and the old woman's body turns into a forked root and her web into a long vine.[23]

The complete entanglement of the Mandan and Hidatsa traditions and ceremonies renders it exceedingly difficult and often impossible to say to which tribe certain features must be ascribed. The Mandans, however, are known to have had for at least one hundred years an origin tradition, of which the following is a good version:

"Many hundreds of years ago the Mandan Corn People came out upon the earth at a place somewhere down along the Missouri River.[24] The exact spot is not known, although the older Indians think that they could find it by tracing the old villages, and certain landmarks.

"These Mandans were really corn people living down under the ground. One day one of these people saw an opening above into this world. Soon all the people began to talk about the world up above, and to wonder about it. So the great chiefs of the corn people sent the black-bird to go up and . . . look through the opening. He returned and told the chiefs that there was a beautiful prosperous country up there which the

[23] Dorsey, *Traditions of the Skidi Pawnee*, p. 36.

[24] Several of the Mandan traditions state that the underground home from which they came was east of the Missouri, which agrees with the statements of the quite similar Arikara tales.

corn people would enjoy. At once the people began agitating the question of going up into this new country. At last the chiefs magically caused a vine to grow up to the small opening. They then sent the fox up to enlarge the hole; after he had done his best they sent up the red fox. The badger was the third animal sent up to enlarge the opening, and fourth and last the elk went up and completed the work with his broad antlers.[25]

After this the people began climbing up one by one and the chiefs led them into the world above. For four days and four nights the people kept steadily climbing out of the hole. Then a woman heavy with child started up and the vine broke with her so that no more people were able to ascend.

Now the Mandan corn people were in the upper world with their chiefs. The head chief was Good Fur Robe, then there were Long Ear-rings, Uses His Head for a Rattle,[26] Swaying Corn Plant, and their sister, Yellow Corn Woman. These five were the leaders of the new nation. Good Fur Robe located and laid out the first village built by these people. This village was laid out with the houses all in rows to represent a field of corn. The same chief also laid out the fields and allotted the fields among the families.

After this Good Fur Robe started the ceremony of cleansing the seed. He had in his possession many varieties of corn, beans, and squashes. The village caller

---

[25] The Arikara versions of this tale are similar to the Mandan one; each version gives a somewhat different account as to which animals helped.

[26] This name evidently refers to the gourd, often used as a rattle.

was first sent out to give notice that the chief would issue seed on the following day. The next morning all the women came to his lodge with presents in their arms to receive the cleansed seed. Each woman was asked what kind of corn she wished, what kind of beans, and what kind of squashes. Then each woman received two seeds of each sort. When they had gotten home the seeds had increased to a quantity sufficient for each one's planting, and the women then went out and planted their fields.

"For a long time these Mandan Corn People lived in this place unmolested. But soon enemies began to bother them and they formed a society for protection, called the Brave Warriors Society. Good Fur Robe said that they must have a symbol, he said that when you take a handful of corn you hold a great deal in your hand, and the same of many sorts of seeds, but when you try to pick up a handful of red beans they are like glass, slippery, and you cannot hang onto them. Therefore he said that they should have a red bean for a symbol to show that they could slip out of the hands of their enemies, and they painted the red beans on their shields. Then he placed two young men ahead of the Brave Warriors and he said that these two would be the defenders of the society and the village. He placed a sunflower stalk in the hand of each one and said that that was to show that they were the bravest of the village and that when they planted the sunflower in the ground in time of war they might never move from that spot even if they must be killed there.[27] This was one of the first societies organized in the tribe."

[27] Old Cheyennes state that one hundred years ago (about 1815) their tribe had a battle with the Mandans on the Upper

The Sun now comes to woo Yellow Corn Woman; she rejects his suit, and in revenge he attempts to destroy the growing corn, but she protects the corn and saves it from him.   In the autumn Good Fur Robe organizes the Goose Society among the women and gives them the power to protect the corn and to see that no one misuses it.   The chiefs and their sister, Yellow Corn Woman, then die and the people put the skulls of the two head chiefs in the sacred bundle.[28]

This Mandan tradition of the origin and adventures of the Corn People reminds one very strongly of the Arikara tale of Mother Corn leading the corn people out of the ground, the main defference being that the part played by Mother Corn in the Arikara tradition is taken by Good Fur Robe and his companions in the Mandan version.   Among the Mandans and Hidatsas we find a "mysterious being" who corresponds closely to Mother Corn in some respects, although in others, and especially in many of the Hidatsa stories, she is an entirely different sort of character.   This is the Old Woman Who Never Dies, or the Grandmother, as she is also called. Many of the Indians say that she belonged originally to the Hidatsas, and it is possible that she was early

---

Missouri, and that a few men on each side "pinned themselves down" to certain spots of ground and remained where they were until killed.   The Cheyenne men who did this belonged to the Dog Soldier Society.   They had "dog ropes" with picket pins attached to them, by means of which they picketted themselves out and remained where they were until they were killed or until their friends carried them off the field by main force.

[28] This custom of placing the head chief's skull in the sacred bundle belonging to his village was a very old one among the Pawnees.

adopted by the Mandans and was given by them some
of the attributes of the Corn Mother. In any event she
is bound up with many of the agricultural beliefs and
practices of the Mandans and Hidatsas. To her many
miracles are ascribed, and frequently these have nothing
to do with agriculture. There is a great volume of
stories about her, most of which have never been collected. The following general account of her is from
Maximilian:

"The Old Woman Who Never Dies has very extensive
plantations of maize, the keepers of which are the great
stag and the white tail stag [the elk and the white tailed
deer?]. She has likewise many blackbirds which help to
guard her property. When she intends to feed these
keepers she summons them and they fall with avidity
upon the maize fields. As these plantations are very
large she requires many laborers and the mouse, the mole,
and the before mentioned stags perform the work. The
birds which fly from the sea shore in the spring represent
the old woman who then travels to the north to visit the
old man who never dies, and who always resides in that
quarter.[29] She does not, however, stop there long, but
generally returns in three or four days. In former times
the old woman's hut was near the Little Missouri where
the Indians often went to visit her. One day twelve
Manitaries came to her and she set before them a pot of

---

[29] Compare Mr. Grinnell's version of the Cheyenne creation
tradition, in the *Journal of American Folk-lore*, v. xx, p. 171.
In this Cheyenne version it is the man who is placed in the south
and the woman in the north. In the spring the man advances
toward the north; he is the Thunder, or is accompanied by the
Thunder, and also by the migratory birds.

maize which was so small that it was not sufficient to satisfy even one, but she invited them to eat and as soon as the pot was empty it was instantly filled and all twelve men had enough. This occurred several times while the old woman resided in that spot.''

Her residence was for a long time on the west side of the Missouri, some ten miles below the Little Missouri River on the banks of a little slough known as the Short Missouri. A single large house-ring here is pointed out as the site of her home and the high bottom there is said to have been the Grandmother's field. According to the traditions she became impatient at the too frequent visits of the Hidatsas and moved into the west.

The Spider Woman of the Pawnees, who kept the corn, the beans, the squashes, and the buffalo in her underground home, resembles the Mandan and Hidatsa Old Woman Who Never Dies, but in some respects she is very different. She lived in the far north, and she had to be forced to give up the seeds and animals which Tirawa through his servants, the Sun and Moon, had given her to distribute among the people. In the Arikara tale of The Girl Who Married a Star,[30] when the girl's son reaches the earth he goes toward the west and finds an old woman who has a large field of corn and squashes and flocks of blackbirds which work for her and which she feeds with corn. She also has many animals. The boy, by means of his magic power, takes the birds and animals (and the corn?) out of the old woman's control and scatters them over the country, so that they may be of use to the people. This is said to have happened

---

[30] Dorsey, *Traditions of the Arikara*, p. 45.

somewhere to the northwest of the Arikara village, which
is said to have been on the Missouri; but the tree which
the girl climbed when she was led to the sky by the star's
magic power, is said to have been near Armstrong.
In this tale, as in the Pawnee versions, the old woman
has to be overpowered and compelled to give up the animals, birds, and corn for the benefit of the people.

Several Cheyenne traditions dealing with the origin of
the corn and buffalo have been collected by Mr. George
Bird Grinnell.[31] There were evidently two sets of these
traditions originally, one set belonging to the Cheyennes
and the other to the Suhtai, a cognate tribe which joined
the Cheyennes long ago north of the Missouri but kept
up a separate tribal organization until about the year
1830.   The traditions of the two people have now be-
some so entangled that it is very difficult, if not impossible, to separate them.   There are two culture heroes:
Sweet Root Standing of the Cheyennes, and Standing
on Ground of the Suhtai, who act together in bringing
the corn and buffalo to the people.   In some of the versions Standing on Ground, the Suhtai, is the leader and
Sweet Root Standing is the follower or servant, but the
positions are reversed in at least one of the tales, and
Sweet Root Standing takes the lead.   Each of these men
has several names.   Sweet Root Standing, more often
called simply Sweet Root, may possibly be a corn name.
His other names are Rustling Leaf, Rustling Corn Leaf,
Corn Leaf, and Sweet Medicine.   Standing on Ground
is also called Erect Horns, Straight Horns, Red Tassel

[31] George Bird Grinnell, ''Some Early Cheyenne Tales,'' part
i, in *Journal of American Folk-lore*, v. xx, pp. 169-194.

is strange that these two Algonquian tribes, the Cheyennes and Arapahoes, who gave up the cultivation of corn so long ago, should have clung to their ears of sacred corn so persistently.

3. Spring, summer, and fall ceremonies

Of the Pawnee rites connected with the planting and cultivation of corn we have no detailed account; but Mr. Dorsey [12] states that the tribe had many interesting ceremonies of this character, both for corn and for pumpkins. The year began in spring when Tirawa first spoke in thunder; then came the ceremonies connected with the consecration of seed, the preparing and planting of the ground. These spring rites had been taught to the first Pawnee head-priest by the four servants of the Evening Star: Wind, Cloud, Lightning, and Thunder, who also taught him prayers to the Evening Star for abundant crops and good hunts. He is said to have direct charge of the distribution of seed to the women in the spring. Besides these seed and planting rites, the Pawnees also had cultivation, or hoeing, rites, and harvest ceremonies, but no detailed account of any of these things is available.[13] At the summer solstice a human sacrifice was made to the Morning Star — a maiden captured from some hostile tribe usually being the victim. This sacrifice is said to have been an agricultural rite, intended perhaps to insure abundant crops. The Skidi Pawnee kept up these human sacrifices until well on into the nineteenth century.

---

[12] *Traditions of the Skidi Pawnee*, p. xvii.

[13] Consult Dorsey, *Traditions of the Skidi*, p. 21; *Bulletin 30*, Bureau of American Ethnology, v. ii, p. 215 and p. 590.

These Pawnee spring and summer rites were not in charge of a single man or group of men. Each village-group had its own sacred bundle to which certain rites appertained, and these rites were performed at the bundle-shrine by priests who had charge of the bundle and its rites. "Each shrine was in charge of an hereditary keeper, but its rituals and ceremonies were in the keeping of a priesthood open to all proper aspirants." [14]

Like the Pawnees, the Arikaras had many rites which were performed at planting-time, during the growing season, and at harvest. These ceremonies centered around the sacred ears of corn, the Bird Case, and the Seven Gourds.

Another Arikara ceremony was that of placing Mother Corn in the Missouri River, that she might return to the gods with the prayers of the people for abundant crops and long life. It is not stated at what season of the year this ceremony was performed, and it does not appear to have been a regular annual rite, but seems to have been performed only on rare occasions. When Mother Corn was thrown into the river, old moccasins belonging to the Arikara children were collected and placed in the bundle with the ear of corn which symbolized Mother Corn. [15]

---

[14] *Bulletin 30*, Bureau of American Ethnology, v. ii, p. 215. In this sacrifice the maiden represented the Evening Star — the First Mother — who was the wife of the Morning Star, and the object of the rite was evidently the refertilizing of Mother Earth. Similar rites with the same object in view were performed at the shrines of agricultural gods and goddesses in ancient times, in Asia, Europe, and Africa. Consult *The Golden Bough*, v. v; worship of Adonis, Osiris, Cybele, and the Syrian goddess Astarte.

[15] Dorsey, *Traditions of the Arikara*, p. 35.

The Arikara priests are said to have had charge of the seed selection and seed consecration rites.

The Cenis, a Caddoan tribe in Texas, had in the seventeenth century a shrine, in the form of a stool, which was used in their agricultural and other ceremonies. "When the corn is ripe, they gather a certain quantity in a maund or basket, which is placed on a sort of seat or stool, dedicated to that use, which they have a great veneration for. The basket, with the corn, being placed on that honored stool, one of the elders holds out his hands over it, and talks a long time; after which the said old man distributes the corn among the women, and no person is allowed to eat of the new corn until eight days after that ceremony. The same ceremonies are used by them in the cultivation of their grain and produce, but particularly of their tobacco, whereof they have a sort which has smaller leaves than ours." [16] The arms and clothing of the young men, when they first went to war, were also placed on this stool and prayed over by an old man.

Among the Omahas in ancient times, the spring opened with the distribution of seed and the singing of the songs of the corn. Of this the sacred legend says: "The stanzas are many. They begin with the gathering of the kernels. The people talk of where they shall plant. Then the men select the land and wherever each man selects he thrusts a pole in the ground to show that now the corn shall be planted." [17] These stanzas and the

[16] Joutel's Journal, 1687, in French, *Louisiana Historical Collections*, v. i, p. 151.

[17] Fletcher and Le Flesche, p. 261.

rites that accompanied them have been forgotten, but it is stated that in the spring the Honga subgens that kept the sacred red ears, gave these ears to the Inkesabe subgens that performed the rites and gave four seeds of the sacred corn to each of the women. Red was among most, if not all, of the Siouan tribes, the color symbolizing abounding life, and the women placed the four red kernels with their seed corn, to fertilize the seed.

Long after the rites of seed distribution had been discontinued, the custom of singing the corn song was kept up. It was sung before the rites of the White Buffalo Hide were performed. The known stanzas of this corn song are given by Fletcher and Le Flesche, on pp. 262-267:

1

O hasten!
  Behold,
With four roots I stand.
  Behold me!

2

O hasten!
  Behold,
With one leaf I stand.
  Behold me!

3

O hasten!
  Behold,
With two leaves I stand.
  Behold me!

4

O hasten!
  Behold,
With three leaves I stand.
  Behold me!

And so on, to seven leaves; then "With one joint I stand. Behold me!" And so on, to seven joints.[18] Then the

---

[18] The "seven leaves," "seven joints," etc., in this song evidently symbolize the seven divisions of the old Omaha-Ponka tribe. An examination of Omaha corn shows that the plants have more than seven leaves and seven joints.

song continues: "With clothing I stand. . . With light, glossy hair I stand. . . With yellow hair I stand. . . With dark hair I stand. . . With light, glossy tassel I stand. . . With pale tassel I stand. . . . With yellow tassel I stand. . . With fruit possessed I stand:

24

O hasten!
  Grasp ye
My fruit as I stand.
  Pluck me!

25

O hasten!
  Roast by a fire
My fruit as I stand.
  Even roast me!

26

O hasten!
Rip from its cob
My fruit as I stand,
And eat me!

As has been stated above, the Omahas also had rites for the insuring of abundant crops, rain-making rites, and rites to protect the corn from wind, hail, insects, and birds; but all knowledge of these rites has been lost.

Hunter, apparently speaking of the Osages, states that they had a green corn festival, and that the elderly women had complete control of all rites connected with the cultivation of corn. The oldest of these women inspected the fields daily, announced the time for the beginning of the green corn season, and controlled the green corn ceremonies.

The corn rites of the Mandans are said to have been instituted by Good Fur Robe and his companions, who led the people out from under the ground. He also organized the Goose Women Society, and first performed the ceremony of distribution of seed and the seed cleansing rites.

The spring ceremonies of the Mandans begin after the ice has gone out of the river and when the geese and ducks have returned.[19] The sacred (?) seed has been prepared before this, by the Corn Priest, in a special (secret ?) rite. He decides on a day, and then on that day, early in the morning, at about the time when the women are returning from the work of clearing and preparing the fields, he ascends to the top of his lodge, begins to shout and sing, announcing that the time has come for the distribution of the sacred seed. Then the women all come to the sacred lodge, each one bearing a present for the priest. The priest performs the ceremony of distribution, placing a few kernels of the sacred corn in the mouth of each woman. The women then return home and mingle the sacred kernels with the seed corn they have prepared for planting. The cleansing ceremony may be requested any year by any woman, who deems it necessary, but the request must be accompanied by a specially valuable present to the priest. Although it might be held in any year, this ceremony seldom occured oftener than once in three or four years, and on these occasions it seems to have taken the place of the regular distribution of sacred seed.

After the request has been made, accompanied by a sufficient present, the priest agrees to perform the ceremony and has an announcer declare the day and hour

---

[19] We here employ the present tense, but the government has put a stop to all of these old Mandan rites and ceremonies.

from the top of the sacred lodge.   On the appointed day the women repair to the sacred lodge immediately after the morning meal.   They take with them all of their seed for the spring planting, in pots and wooden bowls. Inside the lodge they find the priest, his body naked, wearing a headdress of fox skins; he is adorned with sprigs of young sage, fur moccasins are on his feet; his body is painted red and the upper part of his face blue. This is to be his appearance during the whole of the ensuing growing season.   Between the two back posts of the lodge is stretched a map of the world, drawn on skins.   The priest is seated to the left of the fireplace, smoking the sacred pipe of black stone.[20]   This pipe he is to use during the whole growing season.   As the women arrive, they place their pots and bowls of seed in rows before the map; and after this has been done the rest of the people crowd in and fill the lodge.   The rites are performed by the priest, unassisted.   When all is ready he begins to sing and goes through several songs.   He then takes brushes of mint and performs the actual cleansing by brushing over all of the seed, and all of the people as well.   The cleansing is now concluded and the women return home with their seed; but the priest's work has only begun.

---

[20] The sacred pipe of the Arapahoes is also of black stone and is said to represent the Creator.   During the Sun Dance, Mother Earth is fertilized by smoking in this pipe tobacco mingled with black paint, symbolizing the earth, and red paint, symbolizing the life-giving or fertilizing power.

The map of the world referred to above was destroyed in the burning of the lodge of Moves Slowly, the last Mandan Corn Priest, many years ago and was never replaced.

During the whole summer following the cleansing ceremony, the priest must remain indoors except on the occasions when it is his duty to visit the fields. He must
paint himself ceremonially every morning, singing soft-
ly, and he must wear the same things that he had on
during the cleansing rite, and a heavy winter robe, all
summer. It is not permitted him to bathe even in the
hottest weather, and he must not eat berries or any other
fresh food. Should he do so, an early frost will damage
the corn. If a member of his family dies he may mourn
only four days. If a quarrel occurs in his household it
is sure to bring bad fortune upon the village. Once two
women quarrelled in the priest's family, and immediately afterward a man was killed by the Sioux.

When the first silk appears the priest visits the fields;
and on this occasion, if there is a drought, he walks
through the corn, wearing his heavy winter robe and his
ceremonial dress and paint; he carries the sacred gourd
rattle, and as he walks he sings the very sacred dew song.
During the growing season the priest visits the fields four
times, the fourth visit taking place when the corn is ripe.
If he encounters any woman in the fields when he visits
them, she must give him a present. When any woman
finds corn ripe enough to supply seed for the following
year, she brings an ear to the priest. After eating this
corn he has performed all of his tasks, the cleansing season is ended and the priest is free to resume his ordinary
mode of life again. The rites performed during such a
cleansing season were intended to make the corn grow
well, to protect it from drought, hail, wind, and other
enemies. Scattered Corn states that the corn has never

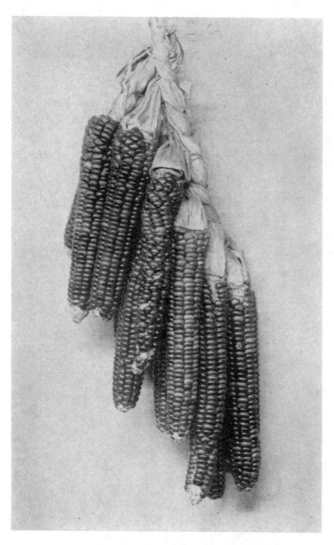

Mandan blue corn

done as well since the agent interdicted the cleansing ceremony. It suffers much more from drought, storms, and early frosts than it formerly did.

The offering to the corn was a harvest-time ceremony. Any woman who feared that her corn crop would be poor might bring a robe to the priest. Upon this robe he marked out a corn plant, with five roots, the leaves, the blossoms, and the ear. The woman outlined the plant in clay and the priest then traced it over. A large cottonwood pole is now brought into the sacred lodge, and the priest wraps the prepared robe about the pole at a height he deems proper. He is then given food; after which there is a general feast. The pole is now borne to the fields (the priest remaining in his lodge), and there it is planted in the ground, and the corn is picked and heaped about the base of the pole. It is supposed that the corn will make a pile high enough to reach the point at which the robe has been wrapped about the pole, no matter how poor the crop may have been before this rite was performed. The priest received as recompense for his work a present for each root and leaf of the corn plant drawn on the robe. The robe after the piling of the corn is over becomes the property of the first person who touches or takes hold of it.

Among the Mandans the ceremonies having to do with the corn come under the heads of ceremonies proper and dances. The dance features belong to the band of Goose Women, and it is only of their activities that we have any account from early travelers. Recent investigation has brought to light, however, a number of important ceremonies and rites which were directly in charge of an hereditary Corn Priest who held his office for life and was a very important person in the community. The last priest, Moves Slowly, who died about ten years ago, was the last of a line of thirty-four priests whose names are kept in a pictographic record, with their ages, which average between sixty and seventy years. This Corn Priest was always the keeper of one of the sacred turtles.[21]

The Goose Women had a regular spring dance and another in the fall. At these ceremonies the Corn Priest acted as drummer, or musician, and was called the Corn Singer. According to Scattered Corn (the daughter of the last Corn Priest) in the ceremonies performed by the priest the Goose Women acted as his helpers and were supposed to be under his direction during the whole of the growing season. It seems probable that the Goose Women features of these ceremonies come primarily from the Hidatsas, along with the regard for the Old Woman Who Never Dies, and that it has been made to fit in with the earlier Mandan ceremonies of the Corn Priest.[22]

According to Scattered Corn the Goose Women might get up dances at any time during the summer, but the most important one was in the fall, after harvest. Maxi-

---

[21] The Arapahoes also keep a sacred Turtle, with the sacred ear of corn, in their tribal medicine-bundle. The Omahas had a group of Turtle people, who appear to have had important rites connected with the cultivation of the soil.

[22] This material, and the following account of the Corn Priest and his duties, with certain features of the Goose Women ceremonies, was collected in recent years among the Mandans by Mr. George F. Will. Scattered Corn was the principal informant.

milian and other early travelers seem to lay more stress
on the spring dance, which they describe at some length.
This dance was held when the wild geese — the messengers of the Old Woman Who Never Dies — first appeared in the spring. The ceremony seems to have been
a short one, as Maximilian states that it was all over by
11 o'clock in the morning, ''but some of the women remained the whole day reclining near the offerings hung
up in the prairie.'' In the following account Maximilian seems to have the Goose Women dance confused
with the Corn Priest's spring ceremonies, part of which
he describes.

Maximilian (p. 334): ''It is a consecration of the
grain to be sown and is called the corn dance feast of
the women. The Old Woman Who Never Dies sends in
the spring the waterfowl, swans, geese, and ducks as
symbols of the kinds of grain cultivated by the Indians.
The wild goose signifies maize, the swan, the gourd; the
duck, the beans. It is the Old Woman that causes the
plants to grow and therefore she sends these birds as her
signs and representatives. It is very seldom that eleven
wild geese are found together in the spring; but if it
happens this is a sign that the crop of maize will be remarkably fine. The Indians keep a large quantity of
dried flesh in readiness for the time in the spring when
the birds arrive, that they may immediately celebrate the
corn feast of the women.''

They hang the meat, together with other articles intended as offerings to the Old Woman, on long racks set
up in the prairie near the village. ''The elderly females
[Goose Women], as representatives of the Old Woman,

assemble on a certain day about the stages [on which the meat is hung up] carrying [each of them] a stick in their hands to one end of which a head [ear] of maize is fastened. Sitting down on a circle they plant their sticks in the ground before them and then dance around the stages.

"Some old men beat the drum, and rattle the schisschikue. The maize is not wetted or sprinkled as many believe, but on the contrary it is supposed that such a practice would be injurious. While the old women perform these ceremonies, the younger ones come and put some dry pulverized meat into their mouths for which each of them receives in turn a grain of the consecrated maize which she eats, three or four grains are put into her dish which are afterwards carefully mixed with the seed to be sown in order to make it thrive and yield an abundant crop. The dried flesh on the stages is the perquisite of the aged females, as the representatives of the Old Woman. During the ceremony it is not unusual for some men of the Band of Dogs to come and pull a large piece of flesh from the poles and carry it off. The members of this band being men of distinction, no opposition can be offered."

Of the autumn dance of the Goose Women, Maximilian says (p. 335): "A similar corn feast is repeated in the autumn but at that season it is held for the purpose of attracting the herds of buffaloes and of obtaining a large supply of meat. Each woman then has not a stick with a head of maize as in the former instance but a whole plant . . . pulled up by the roots. They designate the maize as well as the birds which are the symbols of the fruits of the earth by the name of Old Woman Who Never Dies, and call upon them in the autumn, saying, Mother have pity on us, do not send the severe cold too soon so that we may have a sufficient supply of meat, do not permit all the game to go away so that we may have something for the winter.

"In autumn when the birds emigrate to the south or as the Indians express it, return to the Old Woman, they believe that they take with them presents, especially the dried flesh that was hung up at the entrance of the village for the giver and protectress of the crops. They further imagine that the old woman partakes of the flesh. Some poor females among these Indians, who are not able to offer flesh or any valuable gifts, take a piece of parchment [parfleche — rawhide] in which they wrap the foot of the buffalo and suspend it to one of the poles as their offering. The birds on their return go to the Old Woman each bringing something from the Indians, but toward the end one approaches and says, 'I have very little to give you for I have received only a very mean gift.' To this the Old Woman on receiving the buffalo's foot from the poor women or widows, says, 'That is just what I love, this poor offering is more dear to me than all the other presents however costly. . .' "

Such were some of the innocent beliefs and practices which the United States Indian Office has seen fit to wage a systematic campaign against, until the Indian women no longer dare to send their poor little gifts south with the migrating geese and ducks to the protectress of their fields.

Dr. Lowie's modern account of the Goose Women's

ceremonies is much fuller and more accurate than Maximilian's. His informants stated that before the great
ceremony of the society could be performed it was necessary that someone should have had a dream to that effect.
Then the members prepared dried meat. Calf-woman
says that in the winter some woman would always get
up, saying, "In the spring, when the snow is off the
ground, we are going to have a ceremony, we shall have
to hang up offerings on posts." Then the preparations
were made. When the geese made their first appearance in the spring, meat was suspended from a tripod
meat-rack set up on the borders of the village. When
all was ready, the members of the society paraded
through the village, halting four times on the way to
the meat-rack. Each woman carried on her left arm an
armful of sage, enclosing an ear of corn. Calf-woman
(who was one of the two young girls of the society who
wore duck-skin head-dresses) carried a pipe as well as
some meat and fat impaled on a cottonwood branch.
This pipe and the stick she afterward placed before one
of the male singers, who lit the pipe, seized the dried
meat, and returned it to Calf-woman. This was done
four times. When the procession had arrived at the meat-
racks the Goose Women performed one dance. Then
there came from the village two representatives from
each of the men's societies, in full regalia. These men
were the bravest in their organizations; they approached
the meat afoot or mounted (according to the nature of
their martial exploits) and appropriated the dried meat,
in place of which each warrior left one of his best blankets or a horse for the Goose Woman who had prepared

the food — *i. e.*, the woman who took the initiative in getting up the ceremony. After the performance of the first dance, this woman distributed a great deal of meat to the spectators, whose place was on the west side. After each of the four dances this distribution took place, and after the last dance those who had been newly adopted into the Goose Society gave presents to their adoptive mothers; then each new member took up her sage and corn and raced to the meat-rack and back again. It was believed that the woman who won the race would be instructed by the spirits in the right way of living. After the race the runners cleansed (brushed) themselves with the sage. The singer, to whom Calf-woman had presented the pipe and meat, now turned his robe hair-side out, tied a red-fox skin round his head, and moved with the pipe toward the east, touching what meat remained on the racks with the pipe. This meat was then appropriated by the "mothers" in the Goose Society. All now returned to the village; there a sweat lodge was made, and after all of the women had entered the chief singer (Corn Priest?) also went in. He chanted, dipped some sage in water, and sprinkled all of the women. The "mothers" of the society now prepared food and gave it to their "daughters;" a general feast then followed.

In this ceremony the two middle officers (young girls) wore the duck-bill headband, but the leader and the rear officer wore no distinctive badge. The women on the left side of the lodge painted their faces black between mouth and chin, while those on the right used blue paint. The musicians had drums but no rattles.

The object of the ceremony was to make the corn grow.   The geese and the corn were supposed to be one and the same thing.   It may be noted in this connection that one of the Mandan fieldsongs speaks of the Corn returning to its home "in the east" every autumn.   In the other references to the old woman who keeps the corn it is usually implied that her home is in the south.

During the spring or fall, according to Owl-woman, anyone, man, woman or child, might make "an offering to the geese."   In this ceremony the person making the offering invited the Goose Society to his lodge to feast. The society marched in order to the lodge, the singers going ahead.   They halted and danced four times on the way.   On reaching the lodge the rear officer went in first. One of the singers sprinkled sage for incense near the central fireplace and then all of the members approached and scented their blankets.   They then took their places and the host brought in the calico or other offering for the geese; also presents for the musicians.   The dance then started.   There were four dances and four sets of four songs each.   There was an intermission between the dances and between each set of four songs.   After the dance, one of the singers took a stick and impaled some food on it; this he offered to the four quarters, then threw it into the fireplace.   The host now went to the singers and induced them to utter a prayer in his behalf, asking for long life, prosperity, success in war, etc.   He also went to the members of the Goose Society, and any of them having personal medicines placed a little in his mouth, at the same time praying to the Corn in his be-half.   The offerings made were now distributed among the members of the society and a general feast concluded the ceremony.

The two or three hand-drums used by the Goose Society are said to have been painted with representations of goose tracks.

The autumn ceremony of the Goose Women was more important than the spring dance, according to Scattered Corn. Meat-racks were set up and the women danced four dances, just as in the spring corn dance; but the fall dance was primarily a buffalo ceremony, intended to insure a good fall hunt. Scattered Corn states that one side or half of the Goose Society painted part of the face black, while the rest of the members, those of the other side, painted part of the face blue and white.

At the ceremonial feasts of the Goose Women, Calf Woman, who was one of the two young members who wore duck-skin headbands, always served food to the Corn. She took a piece of meat and offered it to the Corn, saying: "You, Corn, eat this. I pray to you, in order that the members of my society may live long."

The Hidatsas had very few ceremonial observances connected with agriculture, beyond the dances of the Goose Women and the features adopted from the Mandans. Matthews says that this tribe had no important ceremonies connected with corn, while Maximilian states that the corn dance or feast of these people was adopted from the Mandans.

Maximilian describes one rather important agricultural rite which seems to be peculiar to the Hidatsas. Speaking of the beliefs concerning the Old Woman Who Never Dies, he says (p. 373) : "She gave the Manitaries a couple of pots which they still preserve as sacred treasures, and employ as medicine or charms on certain occasions. She directed the ancestors of these Indians to preserve the pots and to remember the great waters . . . from which all animals came cheerfully. . . The red shouldered oriole came at that time out of the water as well as all the other birds which still sing on the banks of the rivers. The Manitaries therefore look on all these birds as medicine for their plantations of maize and attend to their song. At the times when these birds came north in spring they were directed by the Old Woman to fill these pots with water, to be merry to dance and to bathe in order to put them in mind of the great flood. When their fields are threatened with drought they are to celebrate a medicine feast with the old grandmother's pots, in order to beg for rain, this is properly the destination of the pots. The medicine men are still paid on such occasions to sing for four days together in the huts while the pots remain filled with water.''

4.   Various ceremonies, beliefs, and practices

Under this heading we will give some account of the ceremonies, beliefs, and practices in which the corn was brought in but which were not directly connected with the practice of agriculture.

Of the ceremonies belonging to this class by far the most important was that one known to the whites as the Calumet Dance but called by the Pawnees the Hako [23]

---

[23] Hako is the name given to the ceremony by Miss Alice Fletcher. She states, however, that the Pawnee called it Skari, or Many Hands.

and by the Omahas the Wawan. This great ceremony was perhaps of Caddoan origin. Joutel saw it performed in a Caddoan village in Arkansas, 1687, and at an even earlier date it had spread as far as the Sioux and the tribes near the western end of Lake Superior.

The calumet ceremony was primarily a rite of adoption, and the name for the ceremony among some of the western Siouan tribes had the meaning "to make a sacred kinship." A group of persons known as the Fathers gave the ceremony for the benefit of a group of persons known as the Children. In the long series of rites and prayers the Fathers adopted the Children and gave them long life, prosperity, and happiness. Kinsmen might not perform the rites for the benefit of their blood kinsmen, the people of their own gens or clan; but the men of one gens or clan might give the ceremony for the men of another gens or clan in the tribe. More often the ceremony was given by one tribe to another tribe.

In the Pawnee calumet ceremony, or Hako, the sacred ear of white corn played an important part. It was called Mother, and seems to have symbolized the earth mother and her fruitfulness.

Two hollow stems, one male and the other female, were used in the Hako ceremony, and one of the most interesting rites was performed with the female stem. The pipe bowl belonging to a Pawnee Rain Shrine was attached to this female stem by the Rain Priest, who then performed a rite known to him alone — the offering of smoke which symbolized the fertilizing of Mother Earth by Tirawa. At the conclusion of this rite all of the persons who were being adopted as "children" through the

Hako ceremony, smoked the pipe, and by this act they obtained long life, happiness, and the power to beget children. This act of fertilizing Mother Earth by means of the female stem and the Rain Shrine bowl, taken in connection with many peculiar smoke-offering ceremonies practiced by the Indians, especially with the object of procuring abundant food and abundant life, raises the interesting question as to whether smoking was not originally a rite symbolizing the act of fertilization: the breathing of life into all things. It is well known that in early times the Indians never smoked merely for pleasure. The act of taking smoke was a rite; women did not smoke, and we have several statements to the effect that the men grew the tobacco themselves in isolated "tobacco-gardens" although the women grew all of the other crops, with the possible exception of the sacred corn, which, in the case of the Iowas at least, the priests seem to have taken care of.

A curious rite practiced by the Osages, is described by Matthew Clarkson, an English trader in the Illinois country, in his diary, December, 1766.[24] He says: "Mons. Jeredot, the elder, who has been a trader for many years among most of the Indian nations about the River Mississippi, informs me, December 22d, 1766, that the Osages live on a river of the same name, which falls into the Missouri. . . He says that they have a feast which they generally celebrate about the month of March, when they bake a large (corn) cake of about three or four feet in diameter, and of two or three inches in thickness. This is cut into pieces from the center to

[24] Illinois Historical Society *Collections*, xi, p. 359.

the circumference, and the principal chief or warrior arises and advances to the cake, where he declares his valor and recounts his noble actions. If he is not contradicted, or no one has aught to allege against him, he takes a piece of the cake and distributes it among the nation, repeating to them his noble exploits and exhorting them to imitate them. Another then approaches, and in the same manner recounts his achievements, and proceeds as before. Should anyone attempt to take of the cake to whose character there is the least exception, he is stigmatized and set aside as a poltroon.''

This rite reminds one of the well-known ''striking the post'' ceremony formerly practiced by so many tribes living east of the Mississippi.

Mr. John Paul Jones quoted the passage from Clarkson's journal (as given above) in a Kansas City journal years ago,[25] and, commenting on Clarkson's statement, he says: ''Among some tribes on the Missouri River there was a feast celebrated at which the maidens participated, and which resembled the feast of the Osages, except that in the former case it was the character of the maidens for chastity that underwent the ordeal of a challenge.'' This seems to imply that in the maiden's feast a large corn cake was used, as in the Osage warrior rite. In this connection it may be noted that several years ago an old pioneer living near the mouth of Kansas River wrote to one of the present authors and offered to sell him some seed of an old-time variety of very early

[25] ''Early Notices of the Missouri River and Indians,'' by John Paul Jones; Kansas City *Review of Science and Industry,* v. v, no. 5, pp. 286-291, 1881-1882.

yellow corn, called "Maiden Corn." A reply was sent, but the old man failed to send the seed or to answer further inquiries. We might conjecture that this yellow maiden corn was the variety set aside for use in the ceremony Mr. Jones refers to. It would be interesting to know the source of Mr. Jones's information concerning the maidens' feast, and which of the Missouri River tribes practiced this rite. A careful search of available material has failed to disclose any further information of these points.

Maximilian mentions in several places the dancing of the Hidatsa women, accompanied by conjuring tricks. These he terms the medicine dances of the women, and he describes one of them during which a woman pretended to have an ear of corn in her body "which she cast out by dancing." This may have been a practice of Arikara origin. The Arikara were great conjurers and the feat described by Maximilian recalls the Arikara tradition of how Mother Corn was struck by Whirlwind and cast out four ears of corn that were in her body.

Although not exactly a ceremonial feature, the field-songs of the women of the Upper Missouri tribes are very interesting. Many of these songs are remembered and are sung today by the older Mandan and Hidatsa women, and similar songs appear to have been in use at one time among the Omahas, Pawnees, and Arikaras. According to Scattered Corn, some field-songs were the personal property of certain women, who doubtless composed them, while other songs were in general use among the women and were handed down from mother to daughter.

Scattered Corn sang a number of the Mandan field-songs to one of the present authors, and explained their meaning.

The first song, a rather pleasing tune, as Indian music goes, was translated thus:

"My friends, look at the corn; I love the corn. But in the fall when she goes back east, I am sad."

Another of these songs is a sort of lament. The woman who is supposed to be speaking had been married to a very kind husband who had loved her and taken care of her. He was killed while away with a war party and the woman is now married to a man who treats her very badly. She laments her first husband.

A third song is the lament of a young girl for her lover who has gone to war and has fallen after especially distinguishing himself.

Another Mandan song represents a young girl as lamenting: "This young man talked to me out among the sunflowers and I did not answer him. Now he has gone to war and is killed."

There were a great many of these songs, and a large number of them are still in use among the Mandan women. A phonographic collection of them would be very interesting and valuable.

Buffalo Bird Woman mentions the similar songs of the Hidatsa women. She says that the young girls who sat on the elevated platforms in the corn patches at green-corn time and watched for birds, small boys, and other plunderers, passed much of their time in singing and doing porcupine quill work. Two girls usually watched together, but sometimes others came from nearby patches to visit, and three or four would perch on one platform, gossiping and singing "love-boy" songs. Most of the songs sung by the watchers were of this character — "love-boy" or love songs, but others were of a humorous cast. Small boys hung about the corn patches at this season (in the Cherry month, or August) to steal roasting ears, and one of the songs was addressed to these pests:

> "You bad boys, you are all alike!
> Your bow is like a bent basket-hoop;
> Your arrows are fit for nothing but to shoot into
>   the sky;
> You poor boys, you have to run on the prairie bare-
>   foot!"

Another of Buffalo Bird Woman's songs was one which a girl sang to her ee-ku-pa, or chum:

> " 'My ee-ku-pa, what do you wish to see?' you said
>   to me.
> What I wish to see is the corn silk peeping out of
>   the growing ear;
> But what you wish to see is that naughty young
>   man coming!"

This piece of impudence was sung when any young men of the Dog Society happened to pass the corn patches:

> "Young man of the Dog Society, you said to me,
> 'When I go east on a war party, you will hear news
>   of me, how brave I am!'

I have heard news of you!
When the fight was on, you ran and hid;
And you still think you are a brave young man!
Behold, you have joined the Dog Society;
But I call you just plain dog!'' [26]

---

[26] Buffalo Bird Woman's Story, part iii; published in *The Farmer*, St. Paul, Minn., Dec. 16, 1916, pp. 1732-1740.

VIII — VARIETIES

With regard to the direct point of origin of the corn
raised by the Indians of the Upper Missouri, we have no
definite information and hardly any clues. It is generally supposed that corn came into North America by
two routes: part straight up from Mexico into Texas,
the Southwest, and up the Mississippi; and part around
by way of the Caribbean Islands to Florida whence it
was diffused through the Southeast and East. So far
as our material shows, however, there are no distinguishing points between the corn of the Mississippi Valley and
that further east.

We find flint corn, 8-rowed flour corn, 12 or more
rowed flour corn, and dent corn raised as the staple meal
corns by different tribes. The distribution of these sorts
seems to have been governed by the physical characteristics of the corn itself. Flint corn is the earliest and
hardiest type of the four, and will produce a crop with
the minimum of cultivation. Hence it was the favorite
of the least agricultural tribes such as the Chippewas.
It does better under colder atmospheric conditions also
than do the other sorts, and we find that it was the main
type grown by the New England Indians. In fact the
northerly and not exceptionally agricultural Algonquian
stock seems to have been the final champion and devotee
of the flint type. The flints have lost out with the Mis-

souri Valley tribes, there being only one or two sorts raised where five or six flour corns are found.

Next in point of earliness and hardiness come the 8-rowed flour corns. The flour corns were much preferred by the Indians who depended considerably on agriculture for food supply, for the reason that they were much more easily ground and were better as green corn than the flints. The flour corns do better where the heat is more intense than in the woodland area of Minnesota, Wisconsin, and Michigan, and we find them extended down the Missouri and thence east, south of the lakes, and up into New York where the more agricultural Iroquois had many more varieties of flour than of flint corn.

The larger eared, many-rowed flour corn is still more of a warm climate plant, and we find it on the Lower Missouri and across to the east. It merges east of the Mississippi into the dents of the still more southern tribes with which it is very closely allied. It is significant that the later and heavier eared flour corns of the Iroquois came from the south, according to tradition.

The dent corn was raised over the southern area of the United States very generally, and in Mexico as well, clear to or beyond the international boundary, for perfect dent ears have been found in cliff dwellings in our Southwest.

It seems therefore that similar types of corn were raised over all the parts of the continent where agriculture was practised, where similar climatic, physical, and cultural conditions prevailed.

The one exception to this statement is found in the Pueblo region of the Southwest where a very distinctive type of flour corn has been developed. This type, small kerneled and large cobbed, is different in appearance from any of the other flour corns, and has certain important physical differences, adaptations to desert conditions, which have been investigated and described by Mr. G. N. Collins of the U. S. Department of Agriculture. It is almost certain, however, that this type is a purely local development by breeding, a result of the many centuries spent by the Pueblo people under unusual local conditions.

There is no flint corn raised in the Southwest in modern times, but the newly discovered pre-cliff dweller [1] remains show many flints, among them a white flint of 8 rows very similar in appearance to the white flints raised in the North, from the Upper Missouri region to New England. In the early cave or cliff dwellings are also found many ears of the 8-rowed, larger kerneled flour corns, whence probably came the slender, 8-rowed varieties of the less agricultural Navajos.

It will be seen from the foregoing pages that there is but little in the types of corn generally grown by the Indians to show whence they came. With the exception of the local Pueblo types of the Southwest we find the main types generally distributed. Tradition and probability

[1] I have been called into consultation recently by two members of the Peabody Museum staff, who made some rather startling discoveries in the Southwest last year. Samples of the corn found by them have been sent to me for examination. It would appear from these samples that the corn of the ancient people of the Southwest differed very little from the types usually grown by tribes in other parts of the country and familiar to us today; but that during several centuries the corn of the Southwest has developed types that are now peculiar to that region. — George F. Will.

By permission of the Montana Agricultural Experiment Station

1. WINNEBAGO MIXED FLINT
2. MIXED FLINT, LOWER BRULE

point to a southwestern origin for the Pawnee corn, but there is apparently nothing in the types to either prove or disprove the theory.

With the exception of the two unusual color types [2] of the Pueblo area, with the two similar ones reported by the Pawnees, there seem to be a certain number of possible colors and color combinations which are spread pretty generally over the continent.

We have very few records of corn remains being found in old village sites along the Missouri. Dr. Dinsmore found a few charred cobs in eastern Kansas; these were as slender as modern Pawnee cobs and showed 8 rows of kernels. A cob found by Mr. Robert F. Gilder in an ancient village site in eastern Nebraska showed 12 rows of kernels. Will and Spinden found a considerable quantity of cobs and some kernels in a Mandan site near Bismarck, N. D. The cobs were charred; few were over 6 inches long, while many were much shorter. There appeared to be two types of cob, one type slender, of good length, but seldom over 6 inches long, showing straight regular rows of kernels, 8 to 10 rows on each ear, while the other type of cob, short and somewhat thicker than the first, showed invariably 12 rows of kernels, set very irregularly, as is the case with the short ears or nubbins of most varieties of corn. The few kernels found were in very imperfect condition, but they ap-

---

[2] The peculiar color types here referred to are the ''eyed'' corns, grown in the Southwest at the present time and said, by tradition, to have been formerly grown by the Pawnees. One is a white corn with a dark purple spot or ''eye'' on the center of each kernel; the other is described as a blue corn with white ''eyes.''

peared to be also of two types, a small sort, rather long, resembling popcorn, and a large, almost round sort.[3]

The fact that most of the cobs found in village sites along the Missouri show straight regular rows of kernels may be considered as evidence that the Indians selected their seed ears carefully and kept their strains quite pure in early days, and from other evidence we have every reason to believe that this was the case. From practically all of the Upper Missouri tribes we have traditional evidence that the seed ears were selected with the greatest care. Scattered Corn states that each Mandan woman selected her seed while the corn was being husked and placed the seed ears in a special sack. The ears selected for seed were among the ripest and hardest — long ears with straight rows, well filled out to the tip. These Mandan rules for the choosing of good seed ears differ very little from the scientific requirements as given at the present day by corn specialists. The Mandan seed ears were carefully braided and dried by themselves, and were then stored with extra care in a special seed-cache. The Hidatsas, Omahas, Pawnees, and Arikaras also selected the seed ears with great care. Careful women kept two years' supply of seed always on hand.

The Indians possessed a knowledge of the fact that different varieties of corn will cross and mix, and they carefully selected their seed to insure purity of type. In no case did a single family plant more than two or three varieties; in most cases each variety was

[3] Will and Spinden: *The Mandans: Papers of the Peabody Museum*, v. iii, no. 4, pp. 179-180.

grown by a different member of the family and in a separate garden. The older women say that sixty to one hundred yards apart was sufficient in the sheltered bottom lands to prevent any but a very slight mixing, and thus two varieties might be grown at opposite ends of the same garden.

At the present time a very large number of Indian varieties are badly mixed, almost to the extent of being nothing more than the common "squaw corn," as it is called contemptuously by white people. In each tribe, however, a few of the more careful people still preserve pure strains of the old varieties and they can yet be obtained by diligent search and careful inquiry.

Maximilian listed the Mandan corn and vegetables as follows:

"Of maize there are several varieties and colors to which they give different names. The several varieties are:

"White Maize, Yellow Maize, Red Maize, Spotted Maize, Black Maize, Sweet Maize, very hard Yellow Maize, White or Red Striped Maize, very tender Yellow Maize.

"The gourds are yellow, black, striped, blue, long, and thick shelled.

"The beans are likewise of various sorts, small white beans, black, red and spotted beans.

"The sunflower is a large helianthus which seems perfectly to resemble that cultivated in our gardens [in Germany]. It is planted in rows between the maize. There are two or three varieties with red or black and one with smaller seeds. Very nice cakes are made of these seeds.

"The tobacco cultivated by the Mandans, Manitaries, and Arikaras attains no great height, and is suffered to grow up from the seed without having any care bestowed on it" (p. 274).

Scattered Corn also gives a list corresponding with that of Maximilian as far as his goes, but also mentioning several additional sorts. According to her the varieties of corn grown by the Mandans are these:

Soft Yellow Corn; Hard Yellow, or Yellow Flint; White Flint; Soft White; Red Soft Corn; Clay Red Soft Corn; Spotted Corn; Blue Soft Corn; Black Soft Corn; Pink Soft Corn; Yellow-and-Pink Striped Hard Corn, called Society Corn; Wrinkled or Sugar Corn; and "Keika" Corn, of which no definite description could be obtained.

This gives us thirteen varieties. Reviewing the list we find only two flints specifically mentioned, the white and the yellow. We find also two sorts of red corn. The clay red has been tested and proves to be a very distinct and pure type. Mr. Will has recently found the black corn among the Mandans. It is a dark variant of the common Mandan red. By selection from mixed ears, obtained by him from the Indians, a blue corn has been fairly well isolated in two years. Scattered Corn's blue variety therefore may be supposed to have been a pure strain not very long ago.

Of the spotted corn, Scattered Corn says there were two sorts, one in which each kernel was spotted with various colors, the other apparently a regular squaw corn with kernels of different colors.

The pink corn is a soft white with pink shading and follows type sufficiently to rank as a variety.

As to the "Keika" corn, a definite description of this variety could not be obtained, as the Indians could not describe it accurately with the English vocabulary at their command.

The yellow and pink variety is a yellow corn, striped or blazed with red. Scattered Corn says that it belongs primarily to a Ree society, certain members of which still raise it.

The wrinkled corn is the sweet corn; it is of a brownish red color, when ripe and hard, and of a very good flavor when green.

Scattered Corn says that the varieties raised by her family were soft white, soft yellow, and red sweet corn.

Boller visited the Mandans, Hidatsas, and Arikaras some thirty years after Maximilian. The three tribes had been living together for a whole generation, and they had evidently exchanged corn until it was impossible for any but a close observer to distinguish the corn of one tribe from that of the others. Boller makes no attempt at such a distinction but groups the varieties grown by all three tribes together and describes them as Red, Black, Blue, Yellow, Purple, White, and mixed ears showing all of these colors.

Dr. Wilson was informed by an old Hidatsa woman that their favorite varieties in early times were the white and yellow flint, soft white, soft yellow, and sweet corn.

Although the three Fort Berthold tribes are now widely scattered in family groups over the reservation and have no common field, even a short drive about one of the settled portions would convince anyone that a very considerable quantity of old-time corn is still raised by these tribes, indeed a great many bushels of the old varieties are traded every year at the nearest towns for various supplies. At the fair held on the reservation last fall (1915) the display of native corn beautifully braided and hung on a long rack formed a very interesting feature and was the contribution of a comparatively very small number of corn growers although nearly all of the old varieties were included.

With judicious encouragement there is reason to believe that a renewed interest in the growing of their native corn is being aroused among these Indians, and surely no more satisfactory crop could be raised by them. That this corn does not deserve extinction there is no question, for its sturdy qualities and great vitality and resistance to extreme conditions should prove of service in our northern states and in high altitudes where no other corn can grow. It is significant that the early flints developed from these very varieties by the first white corn breeders along the Upper Missouri are even today the highest yielders and the only sure ripening varieties, year in and year out, throughout the dry portions of Montana and western North Dakota. They produce crops of corn in the high mountain region, withstanding cold nights and even light frosts, and have ripened in Norway and northern Russia besides proving of great value in the dry portions of South Africa and Argentina.

Of the corn of the Nebraska Indians we have no early description except in one of Dorsey's Skidi Pawnee traditions, in which it is stated that a Pawnee woman who was a great corn-grower had the following varieties:

White, Yellow, Red, Blue, Blue with white spots, White with black spots, Blue-speckled and Red-speckled corn.[4]

James Murie of the Pawnees, writing in 1914, gave the following list of varieties now grown, or formerly grown, by his tribe:

"They had or have the red, black, yellow and white corn. Then they had the speckled black and red [that is, blue-speckled and red-speckled — the coloring on the blue-speckled ears looks quite black when the corn is hard and dry]. Then they had a yellow corn, between sweet and yellow corn.[5] Then they had a white and sweet corn. They had two kinds of sweet corn, one early and the other late. They also had different kinds of beans, squashes and pumpkins. Also watermelons, and these melons were very small. . . The first four colors of corn I spoke of are ceremonial corn, especially the white, for it is the Mother Corn. A tassel stands upon top as if it were an eagle down feather."

The Pawnee corn of today appears to be as pure as the Mandan corn; most of the varieties show only a slight mixture, and none of them seem to be as badly mixed as most of the Omaha sorts are.

Of the corn of the Omahas, Ponkas, Iowas, Otoes, and Kansa, we have no early descriptions. Sturtevant found a variety of blue corn known as "Omaha" in use as far east as Sibley, Ill., thirty-five years ago. This appears to have been the common Omaha blue flour corn which these Indians still grow. The Omahas informed Dr. Gil-

[4] Dorsey, *Traditions of the Skidi Pawnee*, p. 295.

[5] Perhaps this is the same as the "very tender yellow maize" of Maximilian's Mandan list. He mentions three kinds of Mandan yellow corn: the flour, the flint, and the "very tender."

more that they formerly had dent corn, flint corn, flour corn, sweet corn, and popcorn. ''Of most of these types they have lost the seed since the coming of the white men. The varieties were kept in purity in the old time, as they inform me, by planting in patches at some distance from each other, and you can see that it was necessary to keep the varieties pure for ceremonial reasons, because for instance, if red corn (which was tabu) became generally mixed among the their corn it would make it impossible for one certain gens to touch any of the corn.'' [6]

As might be supposed from their close relationship and intimacy in early times, the Ponkas and Omahas have the same varieties of corn today. Each tribe, however, preserves some varieties which the other appears to have lost.

As far as could be learned the Otoes have only two varieties of corn at the present time, and the Iowas three.

No information at all could be procured concerning the corn of the Kansa, Missouris, and Osages.

In the descriptive list of Indian varieties of corn that is to follow, we have referred only to the number of suckers observed on the plants. In reality the difference in the size and type of the suckers is more important than their number. The production of suckers by the plants of any given variety is a very variable matter, depending upon soil conditions, dry or wet seasons, and close or thin planting. It is very seldom that a series

---

[6] Letter to Geo. E. Hyde, February 18, 1914.

of plants of the same variety will produce the same number of suckers, even under the same conditions.

The corn of the Fort Berthold Indians — Mandans, Hidatsas, and Arikaras — displays suckers in considerable numbers which are almost as large as the main stalk and which in many cases bear a small ear or nubbin. This accounts for the very bushy appearance of these northern Missouri River varieties as compared with those from the region farther south, and also for the larger number of ears produced by each hill.

The varieties from Nebraska seem to have in most cases as many suckers as the northern sorts, but they are very rudimentary, and in no cases are there any signs of ears or even false ear formations on any of them.[7]

The corn of the Red Lake Chippewas very closely resembles that of the Mandans in appearance and habit of growth, so much so indeed as to make it seem probable that this corn is of Mandan origin. We know that the fur-traders took Mandan corn to the Red River posts

[7] These observations were made by Mr. Will, at Bismarck. Last year, 1916, I planted several of the Nebraska varieties here at Omaha, and my notes show that many of the Pawnee, Otoe, Omaha, and Ponka varieties produce two, three, or four suckers, often almost as thick as the main stalk. I found small ears or nubbins on the ends of the side-shoots of several varieties, especially on plants of Omaha gray and Omaha brown, while false ears were very numerous on plants of Ponka sweet corn. A large per cent of the nubbins that are produced on the ends of side-shoots on Omaha brown plants have short tassels springing from the tips. On one Omaha gray plant I picked an ear six inches long from the end of a side-shoot, and this ear had a tassel four inches long attached to its tip. — George E. Hyde.

in early days, and Tanner states (p. 180) that an Ottawa Indian — perhaps one of the many eastern Indians who were in the employ of the British fur companies — was the first to teach corn growing to the western bands of Chippewas, on Red River.[8] The Red Lake Indians were the only western band of Chippewas who grew corn on any considerable scale as late as 1875. One of the agents for the Wisconsin Chippewas states in his report, 1869, that these bands grow "Red River corn," a very inferior variety of low growth, which does not often ripen. This was the only variety they had, and they grew but little of it. At the present day the corn grown by the Chippewas and Winnebagoes of Wisconsin is of a different type from that of the Red Lake Chippewas of Minnesota. It is intermediate between the Red Lake and Mandan corn and the corn of the Nebraska tribes, but is finer leaved than any of these varieties.

The Iroquois corn in habits and in the number of ears per hill closely resembles the Mandan corn. It suckers

---

[8] The earliest mention of the planting of corn by the Chippewas of Minnesota seems to be in Schoolcraft's *Journal of Travel*, 1820, v. i, p. 183, where it is stated that the Chippewas of the head of the Mississippi, and those of Red Lake, plant corn, which ripens early in August. Pike, 1805-1806, states that the Leech Lake and Sandy Lake Chippewas live on meat, fish, wild rice, and roots; he also says that at the fur-post at Sandy Lake 400 bushels of Irish potatoes "and no other vegetables" were raised in 1805, and that at the Leech Lake fur-post there were 500 chests of wild rice. He does not mention corn at all in connection with the Indians on the head of the Mississippi.

Morgan, in Beach, *Indian Miscellany*, v. i, p. 198, states that the Hurons, who are supposed to have come from the southwest, introduced corn into the Georgian Bay region, north of Lake Huron, and that the Cheyennes introduced it to the Red River region.

as much; the suckers very frequently bear small ears or nubbins, and the suckers are comparatively large, in proportion to the main stalks.

The corn from New Mexico and Arizona suckers heavily, but the suckers are not often productive of ears. The bushy appearance of this Southwestern corn is largely due to the many nodes, placed very close together, and each producing leaves. The leaves of the Pueblo corn are also unusually long; they droop and intermingle in a dense mass, adding to the bushy appearance of the plants.

In the following list the varieties enclosed in parentheses, ( ) and not numbered, are those not found by us, but which we have seen mentioned as varieties formerly grown.

LIST OF VARIETIES [9]

Arikaras

1. Red-and-blue Flint. Flint corn from Fort Berthold, similar to Mandan corn in size and habit. Kernels red and blue, with blue predominating, and some few white kernels. Ears about 7 in. long, 8-rowed. At Bismarck, 1916, grew 42 to 72 in. high; ears borne 4 to 13 in. above ground; plants have 2, 3, 4, and even 6 suckers; most plants have 1 ear and 1 nubbin; ripened at Bismarck, 1916, in 90 days.

2. Pink Flint. Similar to Mandan Hard White, but kernels mostly shaded with pink. Flint corn.

3. Ree Dark Red. Flint corn, similar to Ree Pink, but colors more like the Mandan Soft Red. Some question as to whether this and the Ree Pink are real varieties.

4. White Flour. (Perhaps the same as our Mandan Soft White.)

5. White Flint. Evidently the same as our Mandan Hard White.

---

[9] The ripening dates for Bismarck, 1915, are all much longer than normal as the corn was all planted when the season was much advanced.

6. Blue Indian Corn. Flour corn, mixed blue and red. Very bushy growth; height of stalk, 5 to 6 ft.; height of ears from ground 10 to 18 in. Ripened 1916, 136 days. (This variety is from Wyoming, but the seed is said to have come from the Rees long ago when they lived at the mouth of Grand River. The variety is considerably later in ripening than modern Ree corn.)

7. Purple Flour. Dark purple with a few red seeds. (Same as No. 3?)

8. Yellow Flour. Evidently the same as Mandan Yellow Soft.

9. Light Red Flour. Has a yellow cap on each kernel.

10. Dark Blue Flour. (Same as No. 6 of this list?)

11. Dark Red Flour.

Numbers 4 and 5, and 7 to 11, are varieties secured by Dr. Gilmore from the Fort Berthold reservation, under the name "Arikara corn," several years ago. Most of them are evidently the same as the corresponding "Mandan" sorts.

The list of Ree varieties given above must be considered only provisional. The Ree, Mandan, and Hidatsa varieties have been shifted about to such an extent that it seems impossible at the present day to secure a reliable list of Ree sorts. The following notes may prove of interest:

The white farmers of South Dakota still grow a variety of corn known as "Ree corn," and generally supposed to be of Ree origin. It is a mixed flour or squaw corn, the ears somewhat larger than those of the modern "Mandan" flour corn, grown by the Fort Berthold tribes in North Dakota.

The Indian agent at Fort Peck, Montana, reported in 1878 that some of the employes had planted "Ree corn" there, that it had proved very satisfactory and that an Assiniboin man named Long Fox had planted some of the corn also.

Mr. Mooney states that the Pawnees and Rees both had a variety of light blue corn with long slender ears, evidently blue flour corn.

The corn which Mr. Oscar H. Will procured in North Dakota in the early 80's and which he used in his breeding experiments, was a mixed flint, commonly known as Ree or squaw corn, but obtained from an Hidatsa woman.

Hidatsas

The Hidatsas procured their first corn from the Mandans and in later years perhaps adopted some of the Ree sorts. Their favorite varieties are said to have been the Soft White, Hard White, Soft Yellow, and Hard Yellow. They also grew and highly prized the true sweet corn or sugar corn, which they termed "gummy corn."

Mandans

* 12. Soft Yellow. Flour corn. Height of stalk about 3½ ft., ears borne 3 to 12 inches above ground. Suckers heavily, bears leaves on ends of ears and has many false ears. In 1914 ripe and hard in 90 days, at Bismarck. Ears 8-rowed, in a good season 6 to 7 in. long.

* 13. Soft White. Similar to No. 12 in every way except color.

* 14. Soft Red. Flour corn, as near to a pure red as we get, though nearly half of the crop is white or pinkish white. In some individuals the stalk and leaves show the red color, and even the cob is red to the center, pith and all. A trifle earlier and smaller than the two above.

* 15. Hard Yellow. Flint corn, 8-rowed. A very pure strain and said by the Indians to be their earliest variety. Has 2 to 4 suckers; height of stalk 3 to 5 ft.; ears borne 6 to 17 in. above the ground. Ripened at Bismarck, 1916, in 92 days. Has 1 to 2 ears and 1 nubbin per plant.

16. Hard White. Flint corn. Perhaps a heavier yielder than the Hard Yellow and a little later. Ears often shaded with pink.

* 17. Blue Corn. Kernels rather predominatingly flinty; ears rather short. Grows 34 to 50 in. tall; ears borne 2 to 11 in. above ground; 2 to 5 suckers per plant. Each plant has 1 to 2 ears and 1 to 3 nubbins. Ripe, at Bismarck, 1916, in 96 days.

* 18. Spotted Corn or "Buska" (Mandan name). Flint and flour of mixed colors. Grows 40 to 52 in. tall; ears borne 3 to 10 in. above ground; has 1 to 5 suckers (occasionally none), 1 to 2 ears and 1 to 2 nubbins per plant. Ripe at Bismarck, 1916, in 93 days.

* Indicates Mandan varieties mentioned by Maximilian, 1833.

(This is perhaps the variety Sturtevant refers to as being grown by white farmers in the early 80's. It was called ''Mandan corn'' and was spotted, with white, blue and yellow seeds on each ear. ''Came originally from Dakota.'' The Burleigh County Mixed corn, a popular variety among Dakota farmers of the present day, is said to have been of Mandan origin. It is a mixed yellow, white, and red flint. The Fort Peck mixed flint is also said to have been of Mandan origin. It is very much mixed, both flint and flour kernels, white, yellow, red, blue on a single ear. It has yielded 25 to 40 bushels per acre in Montana, where it is the earliest of all varieties tested by the Experiment Station. The Fort Peck corn was of late origin, no corn being grown at Fort Peck as late as 1878.)

19. Clay Red. Flour corn, color dull purplish-red (about like the common purple-red lilac), with bluish tinge on some kernels. White cob, 8-rowed; ears about 6½ in. long. (Scattered Corn gives it in her list as one of the old Mandan varieties.)

20. Sweet Corn. Color red-brown when hard and dry. Ears 4 to 6½ in. long; 10-rowed; white cob with a red ring about the edge of the pith. Very bushy and leafy; a heavy yielder, often 10 or 12 ears to a hill. Gives roasting ears in 50 to 60 days but dries more slowly than the other sorts. Never, or rarely, eaten green by the Indians. Called Wrinkled Corn by Mandans and Gummy Corn by Hidatsas. Grows 36 to 50 in. tall; ears borne 2 to 13 in. above ground; plants have 2 to 4 suckers, 1 to 2 ears and 1 to 2 nubbins. At Bismarck, 1916 (a poor year for corn), some plants of this variety had 2 large ears and 2 nubbins. Ripe and hard, 1916, in 105 days.

(Pink Corn. Included in Scattered Corn's list but not found by us. A variant of the Soft Red?)

(Black Corn. Included in both Scattered Corn's and Maximilian's lists of Mandan varieties. Mr. George F. Will has recently found this corn. It is a very dark variant of the Mandan Red.)

(Society Corn. Given by Scattered Corn, who states it was of Ree origin. Described as having yellow kernels streaked with red. A very large handsome ear of yellow corn striped with red was in the Nebraska Historical Society's collection three years ago, but it was, I think, a Navajo or Pueblo ear. Dr. Gilmore had a few kernels of a Ree variety of red and yellow corn, but he described it as light red with a yellow cap on each kernel. Mr. Biggar of the Bureau of Plant Industry has found this variety among the Fort Totten and Cheyenne Agency Sioux.)

(Keika Corn. One of Scattered Corn's thirteen varieties of Mandan corn. The interpreter could not explain the meaning of the name or describe the corn clearly, as she did not speak English very well.)

(White or Red Striped Corn. Mentioned by Maximilian. This variety has been found among the Pawnees recently, by Dr. Gilmore, but the Mandans, Arikaras, and Hidatsas appear to have lost it. A few seeds of this red striped corn were found in a mixed lot of Refugee Siou. orn from Manitoba.)

Iowas

21. Dark Blue Flour. (Sent by Joseph Springer, an Iowa man.) Very similar in appearance to the Pawnee and Oto dark blue flour corn: 8-rowed; length of ear 8 in. Almost pure, with a few white and blue-speckled kernels on the ear. Stated by Mr. Springer to be the favorite common-purpose corn among the Iowas.

22. Red Flour. This is a flour corn, sacred in the Aruhkwa or Buffalo gens of the Iowas. Mr. Springer, who is an Aruhkwa man, states that he may not eat this corn, as it is tabu. The ear sent has very much the appearance of the Ponka and Pawnee red flour corns. The kernels are of several shades of red, from very dark to a light salmon-red; 8-rowed; ear about 8 in. long, with reddish cob.

23. Brown Flour. A dark brown flour corn, evidently a pure strain. Resembles the Omaha Brown very closely. Kernels dark beaver brown, a few very dark red; cob reddish; ear about 8 in. long; 8-rowed. Said by Joseph Springer to be a sacred variety, and tabu in a certain Iowa gens, which he fails to name in his letter.

Omahas

Although we have grown Omaha varieties both in Nebraska and

in North Dakota for three or four years, the crop has been injured
each time either by drought or by a cold wet growing season, and
we have not been able to grow any very good ears. The large
ears of Omaha corn, described below, were seen in the collection
of the Nebraska Historical Society's Museum (collected by Dr.
Gilmore. Indeed, Dr. Gilmore was the first to collect seed and data
on all of the Omaha varieties).

24. Brown Flour. The ear in the museum collection was 11
or 12 in. long, 2 in. in diameter; color a rich glossy dark brown,
with some bright red kernels. Some ears grown by us had nearly
half of the kernels of this red color; about half of the ears were
pure brown. Cob reddish; height of plant 6 to 8 ft.; ears borne
18 to 36 in. from the ground; 1 to 2 ears per plant; each plant
has 1 to 4 suckers. Ripened in 1916, at Omaha in 120 and at
Bismarck in 126 days. The color of the ears is at first a light
brown, with a yellowish tinge, deepening to a rich dark brown as
the ears mature. This variety was secured by Dr. Gilmore from
Spafford Woodhull, who appears to be the only one in his tribe
who has this corn.

25. White Flour. Appears to be very badly mixed, most ears
showing a large number of colored kernels, mostly blue. Height
of stalk 6 to 8 ft.; height of ears from ground 1 to 2 ft.; ripened,
at Bismarck, 1915, 140 days.

26. Red Flour. Dull red or maroon. Ears mostly 8-rowed;
stalk 5 to 7 ft. high; ears borne 17 to 36 in. above ground;
ripened, 1915, at Bismarck, 136 days. Looks like red-speckled
corn, rather than solid red.

27. Blue Flour. Original ears in museum collection 10 to 12
in. long, 2 in. thick, and showing much mixture, some of the ears
having more white kernels than blue ones. Our plantings proved
the corn to be badly mixed, but few ears being pure blue. This
variety is quite similar to the Rosebud-Sioux blue corn, both in
appearance of ears and in habits of growth. Height of stalk 5 to
7 ft.; has 1 to 2 ears and 2 to 4 suckers; ears borne 1 to 2 ft.
above ground; ripened at Bismarck, 1915, 132 days.

28. Blue and White Flour. Mixed variety with blue and white
kernels. (Considered a variety by the Indians.) Height of stalk

5 to 9 ft.; ears borne 18 to 40 in. above ground; ripened, 1915, 143 days. Some plants have no suckers, others have 1 or 2.

29. Black and White Mixed. (Considered a variety by Indians.) Ear 8-rowed, 9½ in. long. About half the kernels are black flour, and half pearl-white flint; a few kernels of yellow flint, yellow flour, black, brown, blue, etc. Appears badly mixed.

30. Speckled Flour. (Considered a variety by Indians.) Very badly mixed; some white kernels, others blue-speckled, red-speckled, yellow, reddish, etc. Ear 8-rowed, 9 in. long. At Bismarck, 1916, grew 5½ to 7 ft. high; ears borne 13 to 25 in. above ground; 1 to 3 suckers, 1 to 2 ears and an occasional nubbin on each plant. Ripened, 137 days.

31. Gray Flour. (So called by Indians.) Seems to be the same as the Ponka Gray. Ears 8-rowed, 8 to 9 in. long; white kernels very lightly speckled with blue.

32. Black Flour. (Dark blue before the ears are dried. Black after drying.) A pure strain, most ears being pure black. Ears 8-rowed, 9 to 10 in. long. Stalks about 7 ft. high; ears borne 22 to 33 in. above ground; has 2 to 4 suckers on each plant; ripened at Omaha, 1916, 120 days.

(Popcorn. The Omahas say they formerly grew popcorn but have lost the seed.)

(Sweet Corn. The Omahas say they formerly grew sweet corn, but they have long since lost the seed.)

33. Blue-speckled Flour. Some ears of this variety were in the Omaha ''Gray'' corn.

(We have found no trace of the Omaha Sacred Red corn, unless it is No. 26 of this list, which is a dull reddish or maroon color, and looks like red-speckled corn with the speckling laid on so thickly as to produce an effect of solid coloring on most kernels.)

Otoes

After a long search the following two varieties were obtained from this tribe in Oklahoma:

34. White Flour. A pure strain of white flour corn. Ears 8 to 9 in. long; 1⅝ to 2 in. in diameter; the four ears grown and the original ear obtained from the Indians were all 10-rowed.

Planted at Bismarck, 1916, this corn did not mature. Height of stalk 75 to 108 in.; ears borne 33 to 57 in. above ground; each plant has 1 ear and occasionally a nubbin; above half the plants had no suckers, the rest had one sucker.

35. Black Flour. A good strain, evidently pure. About half of the ears are very long and slender — some of them 14 in. long and only 1¼ in. in diameter; all 8-rowed. At Omaha, 1916, plants grew 6 to 7 ft. high; ears 20 to 39 in. above ground; most plants show 1 to 2 suckers. Ripened 119 days.

Pawnees

Most of our seed of the Pawnee varieties appears to be of quite pure strains; none of them shows the bad mixing that characterizes several of the Omaha sorts which we have tested.

36. White Flour. ("Mother Corn.") A good pure strain; ears 8 to 11 in. long and rather slender—all 8-rowed. Stalk 96 to 120 in. tall; ears borne 37 to 52 in. above ground. At Bismarck, 1916, most plants of this variety had 1 or 2 suckers and 1 large ear and 1 nubbin. Did not ripen at Bismarck, 1916. Mentioned in Skidi tradition as one of the oldest Pawnee sorts. James Murie speaks of it as the most venerated of the four sacred varieties.

37. Yellow Flour. A pure strain of yellow flour corn; 8-rowed; ears resemble those of the White Flour; height of stalk 96 to 120 in.; ears borne 23 to 42 in. above ground; each plant has 1 to 4 suckers and 1 to 2 ears. Ripened at Omaha, 1916, 110 days.

38. Yellow Flint. A true yellow flint; color a deep yellow or orange, seemingly quite pure. Ears 7 to 8½ in. long, 10- and 12-rowed. Stalks 7 to 10 ft. high, but varying considerably with soil conditions. At Bismarck, 1916, grew 84 to 108 in. tall; ears borne 40 to 58 in. above ground; each plant has one large ear, occasionally 2; nubbins few in number; some plants have no suckers, others only 1. A few ears ripened at Bismarck, 1916, in about 90 days.

39. Red Flint (?). Flinty corn, mixed colors with red predominating. One ear is 7½ in. long, 8-rowed, reddish cob. Stalks

8 ft. high, ears borne 3½ ft. from ground; ripened at Bismarck, 1915, 140 days.

40. Blue Flour. Blue-black, and a good pure strain. Ears 8½ to 10½ in., 8-rowed. Grows 6 to 8 ft., 2 to 4 suckers on each plant; ears borne 31 to 39 in. from ground; ripened at Omaha, 1916, 110 days.

41. Sweet Corn. Yellow sugar corn; 10- to 16-rowed; 4 to 7 in. long. Grows 5 to 7 ft. high, very leafy and bushy; ears borne 2 to 3 ft. above ground; 1 to 2 suckers, 1 to 2 ears and 1 nubbin on most plants.

42. White and Red striped. Flour corn. Some of the ears are pure white, the rest are striped with red. About half of the ears are 8-rowed, the other half 10-rowed, 7 to 12½ in. long. Stalks 96 to 120 in. tall; ears borne 20 to 51 in. above ground; each plant has 1 or 2 large ears, or 1 ear and 1 nubbin; some plants have no suckers, others have 2, 3, and 4. Ripe at Omaha, 1916, 110 days. Seed obtained by Dr. Gilmore from Mrs. Charles Knifechief.

43. Blue-speckled Flour. A quite pure strain, most ears showing no mixture, but the coloring does not show evenly, some ears being very lightly speckled (like Ponka Gray), others being of almost solid blue-black color. Ears 7 to 9½ inches, all 8-rowed and slender, like most of the Pawnee corn. Stalks 6 to 8 ft. tall; ears borne 1½ to 3 ft. above ground. Did not ripen at Bismarck, 1915. Ripened at Omaha, 1914, 115 days.

44. Popcorn. Obtained by Dr. Gilmore from Mrs. Charles Knifechief. Length of ear about 4 in.; 12-rowed; little round-topped seeds, like Queen's Golden, about 1-16 in. across; colors, pearl-white, yellow, red, and brownish. A very handsome little ear. None of the seed planted has produced a crop. At Bismarck, in 1916, this corn grew 96 to 120 in. tall; ears borne 38 to 65 in. above ground; about half of the plants were without suckers, the rest had 1 sucker; most plants had only 1 ear or 1 nubbin. Did not ripen.

45. Red Flour. The same as our No. 39? Dr. Gilmore had an ear of this corn in 1913. The color was a light red, similar to our No. 39; ear 6 in. long and quite slender. Mr. G. N. Collins

of the Bureau of Plant Industry sent us a packet of seed of this variety, darker red than Dr. Gilmore's ear. He called it ''Pawnee red flour corn.''

(Lixokonkatit or ''Black-eyed-Corn.'' A variety mentioned in the Skidi traditions as grown by the Pawnees in early times. Described as a white corn with black spots on the kernels. Such a variety is now grown by the Navajo, who call it Cudei and consider it sacred. The dots or ''eyes'' on the white kernels are really dark purple.

(Blue with white spots. Mentioned in the same tradition. This corn is perhaps the same as the variety of blue corn with white ''eyes'' that is still grown today by the Hopi.)

(Red-speckled. Mentioned in the same traditions, and also included in James Murie's list of varieties formerly grown by the Pawnees. Not found by us, among the Pawnees, although Dr. Gilmore found it among the Ponkas and Omahas.)

(Black corn. Mentioned in Murie's list. Perhaps the same as our Blue Flour corn, No. 40, which is a blue-black corn, looking quite black in some lights.)

(''Yellow corn between sweet and yellow corn.'' Mentioned by Murie. Perhaps the same as Maximilian's Mandan variety: ''very tender yellow maize.'' Not found by us.)

(''White and sweet corn.'' Mentioned by Murie. His meaning is not clear.)

(Murie also refers to two kinds of real sweet corn, but does not describe them. The Pawnee today have only one sweet corn.)

From the above list it would appear that the Pawnees have today ten varieties of corn and that they formerly had fifteen or more varieties.

Ponkas

(All of these Ponka varieties were collected by Dr. Gilmore.)

46. Red Flour. Rather mixed, but mostly red corn. At Bismarck, 1916, grew 5½ to 10 ft. high; ears borne 9 to 30 in. above ground; part of plants have no suckers, others 1 or 2; most plants have 1 or 2 ears and 1 nubbin. Ripe at Bismarck, 1916, 130 days; in 1915, 134.

47. Sweet Corn. Cream color, or very light yellow, when ripe

and dry. Most ears are 14-rowed; 5½ to 7 in. long. Grows 5
to 7 ft. high, ears borne 18 to 30 in. above ground, and each
plant has 1 to 4 suckers, and 1 or 2 ears. Ripened at Bismarck,
1914, 130 days. At Omaha, 1916, 110 days.

48. Red-speckled Flour. Rather mixed, but mostly red-
speckled. Height of stalk 5 to 7 ft.; ears borne 1 to 2½ ft.
above ground; ripened at Bismarck, 1915, 138 days. Plants have
1, 2, or 3 suckers. Ears are 5½ to 7 in. long, mostly 10-rowed.
We have only examined a few ears. Most of the ears have white
cobs, but the short thick ears (nubbins) have red cobs in most
cases.

49. Gray Flour. This is a flour corn, and seemingly the best
and purest strain the Ponkas have. It is a blue-speckled corn,
with the kernels so lightly speckled as to make most ears appear
light gray or almost white when seen from a little distance. All
8-rowed, ears 7 to 10 in. long; stalks 6 to 7 ft. high; ears borne
20 to 28 in. above the ground; each plant has 3 to 4 suckers.
Ripened at Omaha, 1916, in 100 days. Of 8 ears examined 7 were
absolutely pure and the 8th ear is supposed to have become mixed
with the Omaha Black which was growing near by.

50. Blue-speckled Flour. Procured from the Ponkas by Dr.
Gilmore. Not grown by us. The kernels are much more heavily
speckled than in the case of the Ponka Gray.

Sioux

We have not included any Sioux corn in the above list of Upper Missouri varieties, as the Sioux did not live on the Missouri
until after 1750 and did not practice agriculture in this region to
any large extent until after the year 1850. Most of the varieties
of corn grown by the Sioux at the present time appear to have been
procured by them from other tribes. The Sioux varieties described below have been collected by us during the past two or
three years:

51. Brule Sioux Corn. From Lower Brule Agency, S. D. This
is flint of mixed colors, mostly yellow and white. Long ears, 8 to
12 in., with 8 to 12 rows of kernels. Height of stalks 8 ft.;
height of ears above ground 1 to 3 ft. Often several ears per
stalk; suckers a good deal; ripened, 1914, at Bismarck, 120 days.

52. Speckled Flour. From Standing Rock Agency, S. D. Flour corn, mostly blue-speckled and red-speckled mixed on the same ear. Evidently obtained from some southern tribe (probably Ponkas or Omahas) as it is too late for the Standing Rock region. Grows 6 to 8 ft.; ears borne 1½ to 3 ft. above the ground. Ripened at Bismarck, 1915, in a specially favorable locality, 138 days.

53. Minniconjou corn. Flour corn of mixed colors, 8-rowed. Mostly blue-black and maroon-red, with a very few white kernels. Ear 9½ in. long.

54. Rosebud Sioux Brown. Some ears of this color appeared in plantings of Rosebud Blue Flour corn, and when planted separately the brown corn came true to type. Seems to be very similar to Omaha Brown; some ears are 10½ in. long, and 10-rowed. At Bismarck plants grew 78 to 96 in. high; ears borne 18 to 32 in. above ground; most plants have 2 to 3 suckers, 1 to 2 ears and often 1 nubbin; ripened, 1916, at Bismarck, 120 days. Has reddish cob.

55. Rosebud Blue Flour. Dark blue, 8-rowed flour corn, seemingly much purer than Omaha Blue. Very similar in habit to No. 54.

56. Minneconjou Red Flint. Ears resemble those of Mandan and Ponka red corn. Eight inches long, 8-rowed, red cob. The Minneconjous have a tradition that they obtained seed from the Rees about 60 years ago and that they still grow this Ree corn.

57. Fort Totten, N. D., Sioux corn. It looks like mixed Mandan corn.

Santees: These Minnesota Sioux were the first to take up agriculture on any large scale. They began to grow corn some time prior to the year 1800, on the Mississippi below the mouth of Minnesota River, and in 1862 they had extensive plantations on the upper Minnesota, around Yellow Medicine. Following the uprising of 1862 they were removed to the Missouri, in Dakota, but soon after fled south into northern Nebraska, where most of them still reside. Whether any of the varieties of corn described below were originally brought from the Minnesota region is not known to us.

58. Santee Yellow Flour. Yellow flour corn with a few dark kernels on most ears. The only ear on hand is 10-rowed, 8 in. long. Grows 6 to 9 ft. tall; ears 15 to 30 in. above the ground; plants have 1, 2, or 3 suckers; ripened at Bismarck, 1916, 126 days.

59. Red Mixed. Flour and flint mixed, color also mixed, but red kernels predominating. Resembles the Ponka Red corn, but with white, blue, red-speckled, and blue speckled kernels on most ears. Grows 66 to 76 in. tall; ears 11 to 27 in. above the ground; ripened at Bismarck, 1915, 141 days. Some plants have no suckers, others 1, 2, or 3.

60. White Flour. The ear on hand is mostly white with some dark kernels; 8-rowed; 7½ in. long. Grows 90 to 108 in. tall; ears borne 28 to 39 in. above the ground; plants have 1 or 2 suckers; ripened at Bismarck, 1915, 133 days.

61. Mixed Flint. Described by the Santee agent as variety "A." Flint corn of mixed colors; 8- to 12-rowed, with very long ears. Grows about 7½ ft. tall; ears 1 to 2 ft. above the ground; 1 to 2 ears per stalk; ripened at Bismarck, 1914, 125 days.

62. Mixed Flour. Described by the Santee agent as variety "B." Soft corn, partly dent, of mixed colors; 12- to 16-rowed, ears long and thick. Grows about 8 ft. tall; ears borne 2½ to 4 ft. above ground; 1 to 2 ears per stalk. Ripened at Bismarck, 1914, 130 days. Evidently mixed with modern dent corn.

Refugee-Sioux. These Sioux are Cut Heads and Minnesota Sioux who fled across the Canadian border following the uprising of 1862. Agent McDonald kindly supplied the following information in 1913: "To one of these Sioux belongs the honor of being the first to plant seed in the rich soil of the County of Dennis (Manitoba). Early in the 70's old Gray Faced Bear planted some corn in what is now Mr. Leverington's farm at the elbow of the Pipestone." Mr. McDonald also states that about 1874 Sam Wacanta brought some corn from the Sisseton Sioux, from near Sisseton, S. D., and planted it on the Oak River Reserve, and that in 1875 Harry Hotain, another Oak River Reserve man, brought corn from Fort Totten, N. D. It would therefore appear that the corn grown at the present time by the Refugee-Sioux of the Oak River and Pipestone Reserves is of mixed Sisseton and Fort Totten origin. The Fort Totten corn is evidently of Mandan origin and the Sisseton corn perhaps of Minnesota origin.

63. Refugee-Sioux Mixed. Although of many colors it is clearly but one variety; this was proved by planting different colors in isolated patches, the yield resulting being in all cases mixed in color. This corn and the two Assiniboin varieties are practically the same and have the appearance of mixed Mandan corn acclimated farther north. This corn is probably the earliest in the world. Height of stalk 2 to 3 ft.; ears borne 1 to 4 in. above the ground, and often appearing to be growing right out of the ground instead of on the stalk. Many ears to a hill, all 8-rowed, seldom over 4 in. and sometimes only 1½ in. long. Ripened at Bismarck, 1914, 75 days, and at Omaha, 1916, 71 days.

The Assiniboin mixed corn from Canada, grown at Bismarck, 1916, ripened in 76 days. Grows 27 to 46 in. high; each plant has 1, 2, or 3 suckers, usually 1 ear and 1 nubbin, sometimes 1 ear and 2 nubbins; ears borne 2 to 10 in. above ground.

64. Fort Peck Assiniboin Mixed. Practically the same as No. 65 but a trifle later in ripening. Originally Ree corn? The agent at Fort Peck reported in 1878 that corn was planted there for the first time that year, that Ree corn was the variety grown and that it had proved very satisfactory.

65. Moose Mountain Assiniboin Mixed. Practically the same as No. 63. Agent Cory of the Moose Mountain Reserve, Carlyle, Sask., states that some seed of this corn was obtained from the Sioux of the Pipestone Reserve by an Assiniboin man of the Moose Mountain Reserve several years ago. Mr. Cory has resided in the Moose Mountain region since 1870 and has never seen any other variety of corn planted by these Indians. The corn is known locally as squaw corn. This variety and the Pipestone Sioux corn are, as far as we could learn, the most northerly varieties grown on the continent.

Chippewas

66. Wisconsin Blue. From the Wisconsin Chippewas. Seems to be mostly flint; small ears, 8-rowed. Stalks slender and leafy. The Chippewa and Winnebago corns seem more slender, with finer leaves and stalks and smaller ears than the Upper Missouri corns.

This is perhaps due to adaptation to less windy, and to more shaded and cooler environment. At Bismarck, 1916, this corn grew 60 to 77 in. high; ears borne 12 to 20 in. above ground; some plants have no suckers, others have 1 or 2; each plant has 1 ear and half the plants have 1 nubbin also; ripened in 119 days.

67. Chippewa "Sioux" Corn. From Wisconsin. Mixed flint; 8-rowed, slender but long ears. Stalks grow 66 to 90 in., ears borne 12 to 36 in. above ground; ripened at Bismarck, 1916, 118 days. Said to be of Sioux origin. Most plants have 1, 2, or 3 suckers.

68. Chippewa "Mandan" Corn. From Wisconsin. Mostly a reddish color. At Bismarck, 1915, planted in rich ground it grew exceptionally rank; stalks 5 to 9 ft. tall, ears borne 1 to 3 ft. above ground; did not ripen. In 1916, at Bismarck, grew 50 to 66 in. tall; ears 10 to 14 in. above ground; most plants have 1 to 3 suckers and 1 ear; very few nubbins; ripened in 128 days.

69. Red Lake Flint. From Red Lake Reserve, Minn. Resembles Mandan white flint; many ears shaded with pink or light red; kernels often very broad; all 8-rowed, 5½ to 7 in. long. Grows about 3½ ft. tall; ears 4 to 12 in. above ground; often several ears per stalk; ripened at Bismarck, 1914, 90 days. According to the reports of the agent for 1869, the Red Lake band was the only band of Minnesota Chippewas that planted corn. They had a strip of good soil along the shore of the lake and raised large crops of corn, and had done so for at least thirty years. The Pembina band, on Red River, did not farm. The Leech Lake and Winnibigoshish (the largest Chippewa bands) did not plant at all.

70. Onion Lake Flint. Obtained from Onion Lake, Ontario, 1913. Has exactly the appearance of the Red Lake Flint. It is grown in gardens at Onion Lake as a green corn, although it is extremely heavy and hard flint corn. Rarely ripens in Ontario. Not tested by us.

71. La Pointe Chippewa White Flint. From Wisconsin. Similar to Red Lake and Onion Lake Chippewa flints. Grows about 3½ ft.; ears borne 8 to 12 in. high; ripened at Bismarck, 1914, 85 to 90 days.

Winnebagoes

72. Nebraska Winnebago Flint. Kernels small, flint; colors red, blue, and white mixed on each ear; short slender ears, sometimes only 4-rowed.   Foliage thick and light, stalk unusually thin; grows 6 to 7 ft., ears 15 to 24 in. above ground; ripened at Bismarck, 1914, 110 days. One to 3 ears per stalk. Evidently brought from Minnesota by the Winnebagoes who were shipped to the Upper Missouri with the Santee Sioux after the uprising of 1862. 'This variety and the next one were collected by Dr. Gilmore.

73. Winnebago White Flint. Pure white, 8-rowed flint from Wisconsin. Grows 4 to 6 ft.; very bushy; ears borne 10 to 20 in. on stalk; ripened at Bismarck, 1915, 130 days.

74. Winnebago Blue Flint. Very similar to our No. 66 Chippewa Blue; stalk 5 to 7 ft. tall; ears 19 to 36 in. above ground; ripened at Bismarck, 1915, 141 days. From Wisconsin. Some plants have no suckers, others have 1 or 2.

Iroquois

The following varieties are all from the Iroquois of New York State. They did very well at Bismarck, 1915. In habits of growth they appear more like the varieties usually cultivated by the whites than like the common Indian sorts.

75. Iroquois Hominy Corn. Flint, 8-rowed, ears very long and slender — one ear is 14 in. long and only 1½ in. in diameter. Color pure creamy-white and a heavy yielder. Grows 5 to 8 ft. with ears 1½ to 3 ft. above ground; has 1 to 3 suckers, 1 to 2 ears, and 1 to 2 nubbins; ripened, 1915, 131 days.

76. Iroquois "It's Spotted" Corn. Flint corn of mixed color with some speckled kernels, 8- to 12-rowed, rather short ears. Stalk 4½ to 7 ft., ears borne 8 to 16 in. up; has 1 to 3 suckers; ripened, 1915, 124 days, at Bismarck.

77. Iroquois Soft Red. Flour corn, colors mixed but mostly pinkish, 8-rowed. Stalks 5 to 8 ft.; ears 1 to 2 ft. above ground; ripened, 1915, 132 days. This variety, and No. 76, the Iroquois state, were found "growing wild" in the southern states and were brought home by Seneca war parties. Some plants have no suckers, others one.

78. Iroquois "Tuscarora Short-eared." White flour corn, mostly 12-rowed with broad kernels and very short thick ears. Grows 5 to 6 ft. tall, ears borne 15 to 30 in. up stalk; has 2, 3, and 4 suckers; ripened, 1915, 132 days. Perhaps brought north from North Carolina long ago by the Tuscaroras.

79. Tuscarora Mixed. Mixed flour corn with fairly long, 8-rowed ears. Produces a number of red ears. Grows about 6 ft.; ears 1 to 2½ ft. up stalk; ripened, 1914, 105 days. Evidently of southern origin.

Varieties from the Southwest

80. Navajo Cudei or Sacred Corn. A white flour corn with a purple cap or dot on each kernel—very odd. Very drought resistant. Produces large ears, some 12 in. long, 12- to 16-rowed and very light in weight, kernels round and usually rather small. Plants very leafy and bushy; in a dry year about 3 ft. high with ears borne close to ground; 1 to 2 suckers; in favorable year 60 to 80 in. high, ears 1 to 3 ft. on stalk. A heavy yielder. Ripened at Bismarck, 1914, in heavy late soil, 115 days, and in 1916, 124 days.

81. Navajo Pink-and-White. Flour corn, white with many kernels shaded with a fine shell-pink; 12- to 16-rowed with small round kernels. Its habit is similar to the Cudei; ripened, 1914, 110 days.

82. Navajo White. Similar to above, but pure white.

83. Navajo Red. Taller and not as bushy as the above sorts; long slender ears, 8-rowed, color mixed with red predominating, and red showing on all kernels. Grows 7½ ft., ears borne 1 to 2 ft. high; very heavy yielder, frequently 3 ears on a stalk; ripened, 1914, at Bismarck, 120 days.

84. Navajo "Rosebud" Corn. Mostly flour corn, mixed colors, with long ears, usually 8-rowed. Rather less characteristic than other sorts from the Southwest. Grows 66 to 92 in., ears 16 to 40 in. up stalk; 1 to 2 suckers; a good yielder; ripe at Bismarck, 1914, 110 days.

85. Hopi White. To all appearances the same as Navajo White.

86. Hopi Blue-and-Purple. Flour corn, ears long, with 10 to 16 rows of small round kernels; color dark blue to purple-black,

the later color has purple cobs as in the case of the very dark ears of Mandan Red. General habit of growth similar to the Moqui and Navajo white corns; stalks 5 to 7 ft., ears 1 to 2 ft. on stalks. Good yielder; ripe at Bismarck, 1914, 120 days.

87.   Hopi Pink-and-White. At Bismarck, 1916, grew 60 to 85 in. high; ears borne 10 to 28 in. up; plants have 1 to 2 suckers, 1 to 2 ears, very few nubbins. A few ears ripened in 124 days.

88.   Hopi Blue. Habit quite similar to the above sort.

89.   Zuñi Blue. At Bismarck, 1916, grew 60 to 72 in. tall; ears borne 9 to 18 in. up; plants have 2 to 3 suckers, and usually 1 ear and 1 nubbin; ripened 120 days.

90.   Zuñi White. A little lower in growth than the above sort, and has fewer suckers. A few ears ripened at Bismarck, 1916, in 120 days.

91.   Zuñi Pink-and-White. Not tested.

92.   Zuñi Red-and-Purple. Not tested.

Additional Varieties

93.   Cherokee Mixed Flint. Mostly reddish in color. At Bismarck, 1914, grew 9 ft. high, ears 4 ft. above ground; very heavy stalk with broad leaves; ripe in 130 days.

94.   Shawnee Mixed Flint. Stalk 8 ft. tall; ears 2 to 4 ft. up stalk; did not ripen at Bismarck in 1914 or 1915.

95.   Wichita Yellow Flour. An 8-rowed flour corn, similar to Pawnee Yellow Flour, but rather badly mixed. Grows about 84 to 110 in. tall; ears 21 to 39 in. up; has 1 to 2 suckers, 1 to 2 ears, and few nubbins.

96.   Wichita Black Flour. Seems a fairly pure black or dark blue flour corn, 8-rowed; grows 68 to 99 in. high at Bismarck; ripe, 1915, 156 days. Few suckers, 1 to 2 ears, few nubbins.

97.   Wichita Red Flour. The original ear supplied by George Bent was dark maroon red with a few white seeds. Seems badly mixed. At Bismarck, 1916, grew 90 to 110 in. tall; ears borne 1½ to 4 ft. on stalk; did not ripen. Some plants have no suckers, the others one each.

98.   Sac and Fox Blue Flour. From Iowa. A mixed flour corn, showing white, blue, and blue-speckled kernels on the same ear. At Bismarck, 1916, grew 84 to 108 in. tall; ears borne 26

to 40 in. up; half of plants are without suckers, most of the rest have only one; each plant bears 1 ear; very few nubbins; ripened, a few ears, in 125 days.

99. Iroquois White Popcorn. Similar to White Rice, but with very short ears. At Bismarck, 1916, grew 56 to 72 in. tall; ears borne 8 to 26 in. up; half of plants have no suckers, the rest 1; bears 1 to 2 ears and frequently 1 nubbin; ripe in 100 days.

100. Iroquois Red Popcorn. Similar to above, but of a deep red color. At Bismarck, 1916, ripened in 131 days.

101. Iroquois Blue. Grows 55 to 65 in. tall; ears borne 4 to 16 in. up; plants have 1 to 2 suckers, 1 ear, very few nubbins; ripe, at Bismarck, 1916, in 128 days.

102. Iroquois "Tuscarora White Squaw." White flour corn, ears 8-rowed, long and slender. Height of stalk 6 to 8 ft. Height of ears 2 to 3 ft. Ripened, 1915, 142 days.

103. Iroquois Yellow Flint. Pure yellow flint 8 to 12 rowed. Height of stalk 5 to 8 ft. Height of ears 1½ to 3 ft. Ripened, 1915, 130 days.

104. Iroquois Sweet Puckered. Pure sweet corn, color white. Height of stalk 5 to 7 ft. Height of ears 2 to 3 ft. In unfavorable ground and did not ripen in 1915.

INDEX